Geographies of Postsecularity

This book explores the hopeful possibility that emerging geographies of postsecularity are able to contribute significantly to the understanding of how common life may be shared, and how caring for the common goods of social justice, well-being, equality, solidarity, and respect for difference may be imagined and practiced. Drawing on recent geographic theory to recalibrate ideas of the postsecular public sphere, the authors develop the case for postsecularity as a condition of being that is characterised by practices of receptive generosity, rapprochement between religious and secular ethics, and a hopeful re-enchantment and re-shaping of desire towards common life. The authors highlight the contested formation of ethical subjectivity under neoliberalism and the emergence of postsecularity within this process as an ethically-attuned politics which changes relations between religion and secularity, and animates novel, hopeful imaginations, subjectivities, and praxes as alternatives to neoliberal norms. The spaces and subjectivities of emergent postsecularity are examined through a series of innovative case studies, including food banks, drug and alcohol treatment, refugee humanitarian activism in Calais, homeless participatory art projects, community responses to the Christchurch earthquakes in New Zealand, amongst others. The book also traces the global conditions for postsecularity beyond the Western and predominantly Christian-secular nexus of engagement.

This is a valuable resource for students in several academic disciplines, including geography, sociology, politics, religious studies, international development, and anthropology. It will be of great interest to secular and faith-based practitioners working in religion, spirituality, politics or more widely in public policy, urban planning, and community development.

Paul Cloke is Professor of Human Geography at the University of Exeter. He has longstanding research interests in the geographies of social marginalisation and exclusion, and in the political and ethical responses from Third Sector organisations to issues of social care and justice. His recent books include *Swept Up Lives* (with J May and S Johnsen; Wiley-Blackwell 2010); *FBOs and Exclusion in European Cities* (with J Beaumont; Policy Press 2012); and *Working Faith* (with J Beaumont and A Williams; Paternoster 2013).

Christopher Baker is William Temple Professor of Religion and Public Life at Goldsmiths, University of London, where he co-directs the Faiths and Civil Society Unit. His publications engage theology and religious studies with sociology, sociology of religion, human geography, planning, and policy studies. His edited volume *Postsecular Cities: Space, Theory and Practice* (with J. Beaumont; Continuum 2011) is a much-cited and authoritative text in this field.

Callum Sutherland is a Human Geographer based at the University of Exeter. His published work examines themes of religion, politics, and spiritual activism, with particular regard to Christian praxis and social justice.

Andrew Williams is Lecturer in Human Geography, Cardiff University, and specialises in questions of ethics, welfare, and politics. His latest books include *Working Faith: Faith-Based Organisations and Urban Social Justice* (with P Cloke and J Beaumont; Paternoster 2013) and *Feeding Austerity? Ethical Ambiguity and Political Possibilities in UK Foodbanks* (with L Cherry, P Cloke, and J May; Wiley-Blackwell, forthcoming).

Routledge Research in Place, Space and Politics

Series Editor: Professor Clive Barnett

University of Exeter, UK

This series offers a forum for original and innovative research that explores the changing geographies of political life. The series engages with a series of key debates about innovative political forms and addresses key concepts of political analysis such as scale, territory and public space. It brings into focus emerging interdisciplinary conversations about the spaces through which power is exercised, legitimised, and contested. Titles within the series range from empirical investigations to theoretical engagements and authors comprise of scholars working in overlapping fields including political geography, political theory, development studies, political sociology, international relations, and urban politics.

Un-making Environmental Activism
Beyond Modern/Colonial Binaries in the GMO Controversy
Doerthe Rosenow

The Challenges of Democracy in the War on Terror
The Liberal State before the Advance of Terrorism
Maximiliano E. Korstanje

The Politics of Settler Colonial Spaces
Forging Indigenous Places in Intertwined Worlds
Edited by Nicole Gombay and Marcela Palomino-Schalscha

Direction and Socio-spatial Theory
A Political Economy of Oriented Practice
Matthew G. Hannah

Geographies of Postsecularity
Re-envisioning Politics, Subjectivity and Ethics
Paul Cloke, Christopher Baker, Callum Sutherland and Andrew Williams

For more information about this series, please visit: www.routledge.com/series/PSP

Geographies of Postsecularity

Re-envisioning Politics, Subjectivity and Ethics

**Paul Cloke, Christopher Baker,
Callum Sutherland and
Andrew Williams**

Routledge
Taylor & Francis Group

LONDON AND NEW YORK

First published 2019
by Routledge

2 Park Square, Milton Park, Abingdon, Oxfordshire OX14 4RN
52 Vanderbilt Avenue, New York, NY 10017

Routledge is an imprint of the Taylor & Francis Group, an informa business

First issued in paperback 2020

Copyright © 2019 Paul Cloke, Christopher Baker, Callum Sutherland and
Andrew Williams

The right of Paul Cloke, Christopher Baker, Callum Sutherland and
Andrew Williams to be identified as authors of this work has been asserted
by them in accordance with sections 77 and 78 of the Copyright, Designs
and Patents Act 1988.

All rights reserved. No part of this book may be reprinted or reproduced or
utilised in any form or by any electronic, mechanical, or other means, now
known or hereafter invented, including photocopying and recording, or in
any information storage or retrieval system, without permission in writing
from the publishers.

Notice:
Product or corporate names may be trademarks or registered trademarks,
and are used only for identification and explanation without intent to
infringe.

British Library Cataloguing-in-Publication Data
A catalogue record for this book is available from the British Library

Library of Congress Cataloging-in-Publication Data
A catalog record has been requested for this book

ISBN: 978-1-138-94673-6 (hbk)
ISBN: 978-0-367-66256-1 (pbk)

Typeset in Times New Roman
by Wearset Ltd, Boldon, Tyne and Wear

Contents

Acknowledgements

The authors would like to express their appreciation to Ruth Anderson at Routledge, and the series editor, Clive Barnett, for their encouragement and expertise in helping to bring this manuscript to fruition. We would also like to thank a host of academic friends and colleagues who have participated with us in our journey with ideas about postsecularity. Special thanks here go to: Justin Beaumont, Sean Carter, Richard Gale, Elaine Graham, Derk Harmannij, Julian Holloway, Jo Little, Jon May, Betsy Olson, Mike Pears, and Sam Thomas for their thoughtfulness, engagement, and enthusiasm.

As always, Paul would like to offer his grateful thanks for the wonderful love, encouragement, and support received from his family – Viv, Liz, Will, Chris, Bronnie, Ethan, Evie (and not forgetting Ringo the dog!). Couldn't do this without you! Huge thanks also to close friends and colleagues in Exeter; in particular, Andy and Callum you have added so much to the richness of academic life there over recent years, as well as – alongside Chris – being critically reflexive and hugely supportive partners in this project. I am also really grateful for the friendship and the close-knit musical community of the Week Four Band at Exeter Vineyard Church, and the Nameless People group at YMCA Exeter. For me, music really has been an antidote to the contemporary conditions of neoliberalised university life!

Chris would like to thank Paul, Andy, and Callum for their critical and supportive approach to the production of this volume. To say to have learnt far more than I feel I have contributed is a huge understatement. Thanks to Dilly, Flossie, and Theo for love, support, and understanding. Thanks to dear friends and colleagues who have shaped my thinking in this work, and whose interdisciplinary approach to religion, belief, politics, and policy has been so influential, as well as a ridiculous amount of fun; especially Elaine Graham, Maria Power, Adam Dinham, Beth Crisp, and John Reader. The work and ethos of the William Temple Foundation continues to inspire and challenge, so huge thanks to the WT family, but especially Professor Canon John Atherton (1939–2016) who set me on the road along which I now travel.

Callum would like to thank Terri, Rich, Rach, Euan, Miggy, Andrew, Hannah, Falky, R.R., A.M., M.C., everyone who helped with my PhD research, the Normans, everyone who has passed through Room 385 (particularly Sue and

Paula), Anna, Liz, Bronwyn, James, Dad, and my dear co-authors for their sustaining love and guidance and for improving – both directly and indirectly – my contributions to this book.

Andrew would like to thank all his friends and colleagues at Cardiff University, especially the PhD community who have been a vital source of encouragement along the way. A vast number of people have helped this book to completion, but you will understand that thanking all by name risks inadvertent omissions. To keep this short, I am hugely grateful for Paul, Callum, and Chris – your friendship and wisdom has been invaluable in this project. Special thanks also must go to Andy, Noona, Dan, Helen, Tara, Diana, Rich, Bobby, Mark, Dave, Bex, Matt, Gemma, Ant, Mara, Kieran, Jen, Lyndsey, Jack, Neil, Agatha, Julian, Geoff, Gary, Brian, Günter, Evelyn, Emma, and Lucy – who, in different ways, and at key times, have supported me through the writing process. Finally, I am thankful for my parents – John and Lynda – and my family – Paul, Ellie, Jake, and Iris; Stephen, Nicki, Myla, and Kaira; and the Wisby clan in Swansea, Birmingham and Cardiff. Your love and encouragement have helped see me through.

1   Introduction

1.1 An approach to postsecularity

This is a book about the hopeful possibility that emerging geographies of post-secularity are able to contribute significantly to the understanding of how common life may be shared, and how caring for the common goods of social justice, well-being, equality, solidarity, and respect for difference may be imagined and practiced. Although the religious and the secular are often defined as binary opposites, our discussion in the book explores alternative configurations of these terms. We regard religion to be conditions of being and cultural systems of belief and faith-practice that seek imperfectly to interconnect humanity with the spiritual and the transcendental. We regard the secular as a political project to deny religion a place in the affairs of state; an imperfect social structure designed to limit conflict by privileging universal human rights above any religious demands. In these terms, then, religion is nether cancelled out, nor taken over by an increasingly secularised society. Rather, over time the religious and the secular are becoming co-assembled in interesting new ways.

Over recent years, the notion of postsecularity has emerged across the humanities and social sciences both as a description of the social, cultural, and political re-emergence or new visibility of religion in the urban public sphere (Beaumont and Baker 2011), and as an analytical frame through which to re-examine the co-production of religious and secular domains in ways that depart from the secularisation thesis (see Olson *et al.* 2013). Drawing on formative ideas from Jurgen Habermas and Klaus Eder, the concept of postsecularity reflects both instances of a vigorous continuation of religion in a continually secularising environment, and a more general rise in public consciousness of religious discourse and social action. Despite evidence of a continuing linear movement from 'the relatively religious to the rather secular' (Woodhead 2012, 374), it is now clear that religious identification, belief, and practice continue to be influential, albeit in Western nations often in a more vicarious form involving believing but not belonging (Davie 2015). Contemporary religion, then, concerns myriad and increasingly pluralised sites of subjective and subaltern cultural reproduction as well as more traditional institutions. Rather than focussing on supposed moves from the religious to the secular or vice versa, we seek to shift the emphasis of these debates towards the particular sites, spaces, and practices where diverse religious, humanist, and secular voices come together dialogically and enter into a learning and experimental process in which secular and religious mentalities can be reflexively transformed.

Previously the secularisation thesis (Berger 1967) had suggested the gradual demise of religion as a relevant discourse in the public arena. Habermas (2010) notes how the differentiation of functional social systems during much of the twentieth century resulted in churches and religious communities withdrawing from much of their wider societal intervention, and increasingly confining themselves to their core duties of pastoral care. At the same time, the practice of personal faith also became more individualised, more associated with private pursuit of ritual and dogma than with agitating for wider social responsibility. However, Habermas points to significant areas of social change that have arrested and even reversed some of these privatising trends. First, given the process of Western transformation into post-colonial immigrant societies, the social integration of immigrant cultures has been at least in part bound up in the question of how to achieve tolerant and hospitable coexistence between different religious communities. Second, there is evidence that cultural and social modernisation does not depend on the necessity of depleting the public and personal relevance of religion. Indeed, as we discuss in Chapter 2, the supposed hopelessness of the current post-political age rests in no small measure on a disillusionment about the capacity of the economics, science, technology, and ethics of the neoliberalised secular age to offer any solutions to fundamental issues of inequality, injustice, and commodity fetishism. Third, and partly in the form of a response to these circumstances, Habermas notes that religion has begun to regain influence in a variety of public spheres, most notably as churches and faith-based organisations increasingly assume a public role of 'communities of interpretation', for example, by using their voice to campaign about key issues of social injustice and to speak truth to power in various ways. Accordingly, the previously hushed-up voice of religion is, according to Eder (2006), beginning to be heard again in the public sphere, a turnaround reflected in Berger's (1999) recognition of the counter-secularising forces manifest in desecularisation, and in Casanova's (2011) modifications to the secularisation thesis, acknowledging that despite multiple and diverse secularisations in the West, and multiple and diverse Western modernities, religion remains relevant and influential despite the onward march of other elements of modernist secularisation.

You may well ask, so what? Given a broad presumption amongst the largely secularised academy of social science that religion (typically illustrated by extreme fundamentalist practices) is either an irrelevant cluster of myths and rituals, or indeed a negative source of illiberal attitudes towards violence and social and cultural alterity, the only cause for concern might be that the secularisation of society has not completed the task of privatising religion and stripping it of its public voice. We want to acknowledge at this early stage that some partnerships between the religious and the secular clearly do have a 'dark side' that becomes apparent when strongly conservative religious and political discourses combine to construct political and ethical battlegrounds from which to oppose human rights in areas, for example, relating to sexuality, gender, and welfare (see, for example, Valentine and Waite 2011). Such instances, although appearing to fulfil the criteria of 'postsecular', contribute nothing of value to the inculcation of more hopeful geographies, and merely serve to reinforce a characterisation of religion as being hand-in-glove with neoliberal politics of subject-formation (see, for example, Hackworth 2012), and with a more general politics of disgust (see Inbar *et al.* 2009).

However, in this book we present a rather different, and (in our view) more progressive notion of postsecularity, and as a start, for clarity of argument, we need to be clear what we think postsecularity is *not*: not a universal epochal shift; not a wholesale regime change of entire cities or nations; not a reversal of secularisation; not a return to some kind of pre-secular; not a campaign that equates religion with illiberal moralities. All of these notions appear to us to be too hefty, blunt, and binaric (see Dwyer 2016) to be useful. Rather we envisage postsecularity as a more context-contingent bubbling up of ethical values arising from amalgams of faith-related and secular determination to relate differently to alterity and become active in support of others by going beyond the social bubble of the normal habitus. These ethical values are marked by an explicit 'crossing-over' of religious and secular narratives, practices, and performances that become visible in key geographical expressions of overcoming difference; in certain spaces devoted to care, welfare, justice, and protest, and in certain expressions of dynamic subjectivity characterised by greater degrees of incommonness and heightened care for the common good. It is for these reasons we place a deliberate emphasis on the concept of *postsecularity* – as a condition of being – in preference to specific time-space conceptions of the 'postsecular', and their philosophical justifications wrapped up in 'postsecularism'. The being of postsecularity is conditioned by a co-productive relationship between faith and reason, involving a commitment to solidarity and an openness to difference. It is about doing something together based on an acceptance of the unknowns and unknowables in particular contexts and being open to what could emerge from a mutual action based on ethical negotiation. It can reflect to varying extents both a relaxation of secular suspicion towards spirituality and related re-enchantment, and a willingness to take religious values out into the secular world without being consumed by the fear that in so doing those values will not be diluted or undermined. In these terms, and as we proceed to examine in later chapters, geographies of postsecularity are evident, and can be comprehended, in normative, empirical, and phenomenological registers, reflecting a blurring of sacred and secular spaces and subjectivities through the co-production of hopeful imaginaries, hopeful ethical sensibilities, and hopeful practices.

1.2 Contexts of postsecularity

One of the key distinctions in our approach to postsecularity is that we recognise it to be context-contingent. This not only applies to the geographical diversity of religion, and the consequent careful assertion that what we are examining here takes a particular form in affluent areas of Europe, Canada, and Australasia (although also traceable in different forms elsewhere – see Chapter 6), but also to particular periods of political and material change. One of the principal objections to the idea of the postsecular (see, for example, Kong 2010; Ley 2011; Wilford 2010) is that it simply describes what is already known to have existed over long historical periods. In one sense, this argument is apparent in our introduction so far; secularisation has patently not killed off religion, neither is religious intervention in the public spaces of wider society a new phenomenon. Prochaska (2008), for example, charts the importance of Christian motivation to philanthropy and the politics of social justice in the UK in the nineteenth century, examining the importance of religious associations and benefactors for the delivery of public services prior to the establishment of the welfare state. Many of the organisations whose roots lay in this period – for example, the Salvation Army – have actively continued their public role over subsequent years and remain part of the landscape of contemporary postsecularity. Equally, religious narratives and organisations are evident in the history of counselling and psychotherapy (Bondi 2013), education (Dwyer and Parutis 2012; Watson 2013) and political activism (Marsh 2003; Smith 1996). We would argue that revisiting these spaces through the gaze of postsecularity has the capacity to reveal a more complex picture of assimilation and mutually reflexive transformation of secular and theological ideas than presented elsewhere. However, we do also want to suggest that the bubbling up of spaces and subjectivities of postsecularity in the present day owes much to the way in which contemporary events are delivering particular phenomenologies of need and of societal change which in turn serve to motivate a desire for collaborative activity. Put simply, the subjective conditions of late-capitalism and late-secularism have fundamentally changed. Ward (2009), for example, identifies globalisation with its attendant multiculturalism and insecure patterns of working life, and postmodernity with its espousal of the ironic, the eclectic, and soft forms of hypersubjectivity, as crucial to the assemblage of new kinds of circumstances, including a reanimated embrace of spirituality.

It is important, then, to acknowledge that the context of postsecularity is changing, and that there are aspects of *emergent postsecularity* that underscore the significance of the contemporary empirical moment (see Williams 2015). For example, the form and intensity of religious/secular crossovers have changed significantly through the multifarious realisation of radically plural societies (Molendijk *et al.* 2010). Established sources of secularity and ideologies of secularism have been reconfigured as liberal democratic states enlist diverse religious groups to deliver social cohesion, representation, and 'culturally appropriate' services (Beckford 2012; De Vries 2006). As a result, ethical values are increasingly being constructed through amalgamations of secular, spiritual, and religious frameworks (Bender and Taves 2012). Similar shifts towards postsecularity are also evident in the discourses and practices of international development and humanitarianism (Ager and Ager 2011; Deneulin and Rakodi 2011; Khanum 2012; Mitchell 2017), and in the growth of 'alternative' economic spaces linked to Islamic influence in global political-economic networks (Atia 2012; Pollard and Samers 2007). Further evidence of postsecularity can be found in the pluralistic sensibilities and horizontalist organisation of recent social movements – for example, Occupy Wall Street, Taksim Gezi Park and the Arab Spring (see Cloke *et al.* 2016; Barbato 2012; Dabashi 2012; Mavelli 2012) – all of which have been marked by an explicit 'crossing over' of religious and secular narratives, symbolism, practice, and performance in public space. These trends, events, and circumstances indicate not so much a differentiation of religion from supposedly secular spheres of political, cultural, and economic life (Wilford 2010), but rather how the mutually constitutive dynamics between religious and secular are becoming increasingly visible in the public domain.

Postsecularity in these terms can be represented as an epiphenomenon of its times; an effect of, response to, and resistance against dramatic global and cultural transformations, often illustrated in terms of how poorer communities and societies reach out to religion as a response to the need for reassurance (see, for example, Davis' 2007 account of *Planet of Slums*). Such illustrations, of course, often serve to reinforce the prejudicial regard for hierarchical forms of religion as an expression of existential insecurity (Norris and Inglehart 2004), but we want to argue that religion – and in particular more non-hierarchical forms of spirituality – can equally be viewed as an intrinsically important and valuable cause of affirmative human activity. For example, the propensity for postsecular collaboration has clearly flourished in the landscape of neoliberal governance, as gaps left by shrinking public service provision and the contracting out of service delivery have been filled at least in part by faith-based and other Third Sector organisations. In a recent discussion of food banking in the UK, Cloke *et al.* (2017) suggest that current responses to food insecurity and poverty is occurring 'in the meantime' – gesturing both to the meanness of neoliberal politics of austerity that disproportionately penalise the poorest members of society, and to the necessity to take immediate action whilst at the same time mobilising an ethics and politics of social justice in resistance to the causes of this poverty. It is the phenomenology of need, coupled with a latent ethical sensibility to act (often in this case founded on theological as well as ideological properties) that may well be causing a wider conscientisation of staff and volunteers in food banks, and a host of other settings of care and welfare. As the welfare state becomes denuded and hollowed out, so a small multitude of people are being prompted to act because of personal and societal experience of the unmet needs of others. Some commentators translate these context-contingent causes and effects as surrender of religious specificity and incorporation into the political ethos of state-led governmentality. Third Sector involvements in welfare are therefore typically interpreted as being co-opted by and attuned to the objectives and values of contemporary governance. Woodhead (2012, 15) gives us one such narrative drawing on recent history when the political left ruled the urban political roost:

Once the churches had thrown in their lot with the welfare state and with secular priorities, however, their distinctiveness was in danger. They became part of the social fabric and the reigning moral and cultural ethos. This was one reason why religion became increasingly invisible in the welfare era. Another was that, once the churches had surrendered control to the state, the partnership could easily be forgotten, particularly by the political left.

We acknowledge that negative public response to faith-based organisations seems to have eased over intervening years (Beaumont and Cloke 2012). However, more generally, such analyses seemingly present an interpretative frame that offers an unhelpful choice when analysing religious public action between being understood as co-option or as resistance; and in so doing obscures some of the more progressive possibilities that can arise in and through the spaces of postsecular action. As we discuss in Chapters 3 and 4, many of these spaces of postsecularity may be more fruitfully understood in terms of a theoretical 'messy middle' (May and Cloke 2014) in which obsession with either/or frameworks of understanding makes it is easy to pass over the ordinary but significant ethics and politics of possibility constructed and performed therein both as an effect of the context, and as a cause of context-specific agency.

If specific temporal phases of globalisation and neoliberal austerity offer one set of political landscapes in which to understand context-contingent postsecularity, then another significant form of context is pedagogic in nature. In short, the heavily secularised nature of the social science academy has often resulted in an unwillingness to recognise religion as a force for good, or as a useful partner in secular endeavours. This secular social scientific gaze has resulted in a reluctance to contemplate the possibility of postsecularity at work, which in turn presents a pedagogic stumbling block to the recognition of any potential hopefulness arising in spaces and subjectivities of postsecularity. If a hegemonic pedagogic interpretation of religion only allows us to interpret faith-based activities as self-serving acts of charity, that at best provide an outlet for liberal guilt and morality, and at worst provide cover for proselytising and entrapment of vulnerable citizens, then it follows automatically that no good can come of such activities, and any scholarship that suggests otherwise is simply uncritical. It is only as this blinkered set of assumptions has been challenged that the recognition and critical examination of geographies of postsecularity has been enabled. A brief review of geographies of religion (see, for example, Hopkins *et al.* 2013) illustrates the rising importance of this challenge. Until very recently, geographical study of religion has been carried out in a marginalised subfield that has struggled to establish itself as mainstream and has been neglected as a source of interdisciplinary or cross-disciplinary initiative (Ley and Tse 2013; Tse 2014). Religion has been the last great otherness in geography; that which has been most shunned and swerved around by the general practice of the subject, which on the whole remains resolutely secular in nature and cautious about conceiving of religion as being interconnected with progressive ethics or politics (see Cloke 2010; 2011). In Yorgason and della Dora's (2009, 629) terms, religion has been the 'terra incognita' of human geography. More recently, however, interesting cross-overs between religion and other geographical issues – gender, mobility, identity, welfare provision, ethicality, and the like – have prompted a gentle repositioning of scholarship on religion and spirituality (see, for example, Bartolini *et al.* 2016; Hopkins *et al.* 2012; Holloway and Valins 2010; Kong 2010; Yorgason and della Dora 2009). As part of this steer – one can hardly think of it as a 'turn' – geographers have become more actively involved in multidisciplinary discussions (sparked initially by the work of Justin Beaumont 2008a; 2008b; Beaumont and Dias 2008) relating to the possibilities of faith-based involvement in wider practices of postsecularity. A series of seminal conferences and edited collections have followed (including Molendijk *et al.* 2010; Beaumont and Baker 2011; Beaumont and Cloke 2012; Gorski *et al.* 2012; Nynas *et al.* 2015) that have served to open up discussion of how geography might respond to and interact with the concept of postsecularity.

One crucial element of these multidisciplinary discussions has been a gradual, and perhaps sometimes grudging, acknowledgement that intellectual appreciation of faith, belief, and religion has needed to change. The starting point in wider social science was similar to that in geography; there were vested interests in the continuing adhesion to the secularisation thesis, not least so as to enable critique of religion as an integral part of the broader Enlightenment agenda. Even those who were responsible for subdisciplines that addressed religion (for example, in sociology and geography) seem to have been reluctant to embrace new ideas about postsecularity, claiming that there was nothing new in such ideas, and that they amounted to a red herring which demonstrated a distinct lack of appreciation for the scholarship that had gone before (see, for example, Beckford 2012; Calhoun *et al.* 2011; Kong 2010; Ley 2010). Abandoning secularisation as the overarching explanatory framework was too big a sacrifice for many (see, for example, Sweeney 2008). However, as reflections on postmodernity began to be taken seriously in social science, a challenge was presented, taking up the baton passed on by Bauman (1992), to wrestle with the issue of how to take difference seriously, and in so doing to rediscover the potential enchantment inherent in explorations of otherness. Such exploration included a re-evaluation of the assumed boundaries of key categories of social scientific endeavour, including secular/religious divisions.

Social science responses to this question, of obvious relevance to geographers, have involved incorporating the increasing visibility of religion into existing social theory, identifying the return of religion as a reaction to the times; hence Castells' (1998) conception of the return of (fundamentalist) religion as a political force – but hardly a progressive one – and Davis' (2007) account of the importance of fundamentalist Pentecostal religion to slum dwellers as a response to their political and social marginalisation. However, the idea that religion may offer other kinds of potentials to society became more established in social science via engagement with Habermas. We detail the key ideas inherent in this engagement in Chapter 2, but one breakthrough notion was Habermas' recognition that religion represented a reservoir of *cultural* autonomy, with its pool of imaginary distinctiveness, that included moral and spiritual resources that could be a significant factor in the renewal of the social contract. In so doing, he broached the possibility that the secular and the religious could be regarded as more equal partners in a more open-ended process of knowledge production, and thus created a peg on which to hang the further possibility that different understandings of religion may have a role to play in the process of re-enchantment.

As the secularisation thesis became more open to discussion, so other leading cultural and social theorists began to reference a postsecular condition (see, for example, Derrida 1998; Taylor 2007; Vattimo 2003 and Žižek 2001) and others (for example, Agamben 2005; Badiou 2003; Eagleton 2010) more specifically turned to religious discourse for new imaginaries about political intervention and radicalism. As Ward (2009, 131) comments, 'it is at this point, the point where religion has a public voice, that religion becomes political again', and it is important to note that this public voice became enabled via the production of new social science knowledge as well as being heard from religious and faith-based organisations themselves. As Ward further argues, the cultural reassertion in religious and other spiritualities of powerful mythic and mystical modes of thought, invoking a re-enchantment in senses of mystery and wonder, contributed both to a greater acknowledgement of the possibility of the sacred, and to a platform of dissatisfaction with, and critique of, neoliberalised secularity:

> This cultural reassertion is the greatest single source of the desecularisation and resacralisation of the West. It is related to … a more general re-enchantment of the real, a return to the mythic and the supernatural, a hastening dematerialisation, the increasing virtuality of the real, and the deepening mystification by many people about the complex scientific workings of quite ordinary things …
>
> (Ward 2009, 147)

Again, we need to emphasise a potential dark side to this assertion about deepening mystification; harmful religious delusions can sometimes be used to stabilise a threatened worldview without substantiation and in ways that produce deleterious effects of othering (see McIntosh and Carmichael 2016).

Interest in postsecularity, then, has emerged out of the nexus of contemporaneously recognising the possibilities for re-enchantment in the spiritual nature of religion and questioning the purported sufficiency of secularity. Despite the difficulties raised for some scholars by any use of the prefix 'post' (and the most difficult cheques/checks always seem to be 'in the post'), postsecularity, as McLennan (2007) observes, is neither built upon an intrinsic anti-secularism nor purports to suggest what comes after or instead of secularism. Rather, it serves as a heuristic conceptual device to question and probe the underlying assumptions of secularity, and in so doing to re-interrogate the faith-reason binary by recognising new modes of belief, new conditions for enactment of belief, and new ways in which the secular and the sacred may be becoming blurred. In short, postsecularity enables a critical engagement with the ways in which the boundless mystery and bounded structure of faith and reason can collaborate in the co-production of more hopeful spaces and subjectivities.

Throughout the book, we examine these engagements and co-productions in considerable detail, but at this stage we want to signpost three particular currents that in our view have become very significant in putting flesh onto the bones of postsecularity: the receptive generosity (Coles 1997) necessary for social movements to perform ethics and politics appropriate to the underlying direction of postsecularity; the partnerships of rapprochement (Cloke and Beaumont 2013) that embody the values and potentials of postsecularity; and the pre-formative ethics of postsecularity that connect with the possibility of reconfigured desire and 'post-disenchantment' (Rose 2017).

1.3 Three currents of postsecularity

The ethics and politics of postsecular caritas

The first of these currents draws on the writings of political theorist Romand Coles (1997 and 2001; Hauerwas and Coles 2010) who explores the possibilities and practices of motivating a more radical sense of generosity within the social movements of radical democracy. Coles argues that ethical relations should be characterised and animated by a deliberately receptive form of generosity involving both an openness to the being and voices of others, as well as a desire to give them something of value:

> The question involves a partly agonistic, partly co-operative – always transfiguring – dialogical effort with others to discern what is lower and what is higher; to discern how these differences and distances might be brought together and held apart such that we might become more receptive of their gifts, more capable of giving, less resentful and revenge-seeking, more radiant. This entwinement of giving and receiving is the precarious elaborating foundation of well-being and sense.
>
> (Coles 1997, 22)

Unless generosity is fashioned in the context of a radical receptivity, he argues, its outcomes will fall short of that which was intended and will be prone to the kinds of violence, imperialism, and assimilation that he suggests have pervaded aspects of both religious and liberal activity in the public sphere. It is notable here that in claiming that overly strict boundaries have tended in the past to limit the numbers and characteristics of people who are able to take part in social movements, Coles is willing to embrace the radicalism located within the discourses of Christian religion, and is open to the possibility – theoretically, and in the agonistic and dialogic character of particular individuals and groups of human beings – that radical politics and radical ecclesia can collaborate in the textures of caring for others. In this way, he is content to mobilise the Christian values of 'caritas' (giving) and 'agape' (sacrificial love), but seeks to transfigure them, and notably the ideal of caritas, arguing for a wider sense of generosity in which no theological or secular position can claim absolute privilege for itself.

He therefore delineates a *postsecular caritas* that seeks transformation through attentive listening, relationship-building, and careful tending to places, common goods, and diverse possibilities for flourishing.

Two particular values emerge from this vision of postsecular caritas. First, it seeks to shift political and theological imaginations beyond contemporary political formations, charting paths beyond the current political economy of endless growth and concentrated power that Coles sees as 'waging war on people and on our planet' (p. 218). This in turn poses pressing questions about exactly how social movements can mobilise their nature and possibility towards a communicative rationalisation of particular aspects of life, and in particular how they focus on the co-production of in-commonness that permit the concerns and engagements of receptive generosity to achieve greater depth and breadth of influence. In other words, how can ethical principles of postsecular caritas be marshalled in a morality of thinking as well as of doing, and how can self-other relations transcend enlightened self-interest and discover an ethics that cultivates both giving beyond equivalence and the truncation of revenge as signified by grace? Such principles require the cultivation of strong ethical sources, both ideological and theological. Second, how can postsecular caritas work across religious and secular boundaries, given that both territories will be required to forgo a privileging of their own position in order to sign up to the pursuit of transformative practices of attentiveness to in-commonness rather than tribal self-interest? The possibilities of postsecularity seem to rest on this capacity for this kind of mutual and receptive generosity in the practices that align social movements consisting of mixed foundational and motivational claims.

Some pointers to these values and questions emerge from a published exchange – *Conversations between a Radical Democrat and a Christian* – between Coles (self-identifying as a member of no church) and Stanley Hauerwas a prominent public intellectual, theologian, and ethicist (Hauerwas and Coles 2010 – see also Hauerwas 1983 and 2000), in which the potential practicalities of a radical politics and ethics that 'goes beyond' secular and religious tribal allegiances are discussed in some detail. In the conversation, Coles is preoccupied by the question of how radical democracy can be exemplified by urban organising practices that engage a wide spectrum of people and bridge over political divisions. It is these local politics that in his view provide an ethical learning ground for receptive generosity that might then potentially infuse political work at other scales. Drawing on the work of the late John Howard Yoder, a Mennonite theologian who championed non-violent social activism (see, for example, Yoder 1994) – but also note his subsequent demise – (see Cramer *et al.* 2014), Coles acknowledges that religious dialogical communities can sometimes model powerful practices of generous solidarity, both in the creative use of conflict, and by being vulnerably receptive to marginalised people in and beyond the church. In this acceptance of some Christian ethical values and practices, Coles translates and develops an 'alien' religious discourse into his own idiom; but his transformation of religious ideas into his own secular frame is accompanied by a clear recognition that the characteristics that he finds admirable in Yoder's patient commitment to non-violence are inseparable from Yoder's Christocentric understanding of Christian discipleship. Postsecular caritas in this instance represents: generously recognising commendable ideas and practices from within alien belief-sets and being willing to use them across religious-secular boundaries; acknowledging the interconnections between ideas, practices, and beliefs, even if these beliefs are not shared; and being impressed that religious groups can be sufficiently confident in the dialogical process of confrontation and reconciliation to realise that they have got things wrong in the past and can learn from secular critique. In turn, Coles recognises the growth of (albeit wary and sometimes distrustful) alliances between secular movements and theological ideas:

> What we see in a number of insurgent struggles is a radical-democratic tradition that is (becoming) distinct from Christian tradition and yet cultivating many proximate virtues and communities of character. Perhaps Bob Moses suggests this more haunting possibility, as may Myles Horton, Howard Zinn, Audrey Lorde, Judith Butler, Larry Goodwyn, Adrienne Rich, Tom Hayden, Saul Alinsky (somewhat), Charles Payne and a plethora of emergent radical-democratic communities reverberating across time. The haunting possibility is that there are traditions of radical-democratic practice that are arising, which, though indebted to several theological practices and visions, are developing admirable and possibly enduring capacities for seeing and moving across the world.
>
> (Hauerwas and Coles 2007, 36)

This 'haunting possibility' of postsecular caritas at work in political movements rests on the possibility of a syncretic radical-democratic community that exceeds what was previously embodied by Christian and non-Christian ethics prior to the collaboration. The generative fusion of Christian and democratic traditions co-produces in Coles' mind more admirable imaginaries and practices than those inherited from more separate religious and secular positionings.

This generous expression of possibility, however, does not come without its worries. Clearly the process of being open to the being and voices of others is often accompanied by the kind of tensions, questions, doubts, and lack of trust that stem from asymmetries of power. As Coles asks: 'Might not ethical practice hinge very significantly on slackening the will to retain identity? Should Christians and radical-democrats have confidence that we have the far larger story and that our task is to outnarrate all the others?' (Hauerwas and Coles 2007, 42). Clearly this concern over an incapacity to tell big narratives without claiming particular identification for one side or other of the religious/secular divide is an issue that applies to all parties in a collaboration. However, the tendency for some religious movements to want to 'badge' social activities in ways that make them exclusive (see Cloke and Pears 2016a; 2016b) tends to legitimate the fear that faith-organisations may now be more adept than others at excluding alternative traditions and narratives from which they have a lot to learn. Coles' discussion of postsecular caritas, then, charts significant landscapes of possibility for the co-production of postsecularity, but at the same time clearly identifies the need that syncretic social movements require a radical notion of insufficiency to underpin the collective receptive generosity that underpins such co-production.

Postsecularity, rapprochement, and reterritorialisation

Alongside Coles' recognition of the possibility of postsecular caritas in emergent social movements, geographers have also emphasised how the context of neoliberal austerity, with its shrinkage of the welfare state, has established a fertile landscape for the propagation of partnerships of religious and secular individuals and organisations seeking to step into the gap in order to meet the needs of marginalised and excluded people. Jack Caputo (2001) illustrates the empirical significance of religiously-motivated social action in the contemporary city:

> If, on any given day, you go into the worst neighbourhoods of the inner cities of most large urban centers, the people you will find there serving the poor and needy, expending their lives and considerable talents attending to the least among us, will almost certainly be religious people – evangelicals and Pentecostalists, social workers with deeply held religious convictions, Christian, Jewish and Islamic, men and women, priests and nuns, black and white. They are the better angels of our nature. They are down in the trenches, out on the streets, serving the widow, the orphan and the stranger…
>
> (p. 92)

Certainly, within a Christianised Western framework, research has shown how faith-based organisations (FBOs) have become important in their own right as significant providers of non-statutory services of care and welfare, but also how these organisations have opened up new possibilities for wider involvement in ethical practices of postsecularity (see Beaumont and Cloke 2012). The specific contribution of FBOs has been clearly charted. The enduring significance of religious faith in the work of community mobilisation in deprived urban communities has meant that social, religious, and spiritual capital has been made available in particular localities via the spaces, time, organisational potential, and ethical motivation of faith-involvement (Baker and Skinner 2006). Moreover, FBOs have been shown to be frequent partners in progressive multi-organisational coalitions working for justice and against poverty. It is in the co-production of such partnership, in *rapprochement* (Cloke and Beaumont 2013), that ideas about spaces and subjectivities have found particular expression. Clearly some FBOs function primarily to support their own faith-networks, but others have been able to transcend these faith-boundaries in a range of different ways. Some consciously pursue a policy of professionalisation, sometimes subjugating their faith ethos to the wider objective of being recognised by secular clients and funders as well-trained, efficient, and open-to-all service-providers in a particular sector. In such cases (such as the Grooms-Shaftesbury organisation relaunched as 'Livability' and the Nationwide Festival of Light rebranded as 'Care') faith-ethos can be kept in the background, so as to enable secular as well as faith-motivated support. In other contexts, FBOs enter into deliberate partnership with secular and religious others to form more avowedly postsecular liaisons (such in the case of London Citizens – see Jamoul and Wills 2008). For such liaisons to be successful, partners need to agree on particular crossover narratives (see Chapter 2) in order to do something about the plight of socially excluded people in the city, and this may mean leaving divergent issues at the door in order to achieve the desired rapprochement.

Some geographical discussion of these phenomena has sought to temper their apparent significance (see, for example, Kong 2010; Lancione 2014; Wilford 2010). Seen in the context of longstanding attentiveness to both sacred and non-sacred spaces of religion, postsecular rapprochement can be interpreted as just the latest phase of a long-running attention to religion in the public sphere. Seen from the perspective of preserving geography as a critical secular space, ideas about rapprochement and postsecularity can be read as an attempt to introduce a normative defence of religion and religious values, and/or as an apologetic for joining in with neoliberal subject-formation. However, an alternative perspective is to be more sensitive (theoretically, epistemologically, empirically) to the expressions, organisations, and practices of postsecularity that may otherwise be masked by the ideological assumptions and driving forces of a universal and linear secularisation thesis. In other words, geographies of postsecular rapprochement may represent a radical departure in the understanding of contemporary society and space in which we refuse to be blind to partnerships involving religious praxis that we might otherwise ignore, or just assume do not exist. As Cloke and Beaumont (2013, 32) have argued:

> It is our contention that there is potential within postsecular rapprochement to embody both an expression of resistance to prevailing injustices under neoliberal global capitalism, and an energy and hope in something that brings more justice for all citizens of our cities rather than simply rewarding the privileged few.

The argument here is for an understanding of the underlying conditions for potential reterritorialisation that may result from engagements with postsecularity. A more detailed examination of these underlying conditions is presented in Chapter 2, but it is worth noting at this stage the growing body of evidence suggesting that actual and potential activities of postsecularity are actively contributing to the reterritorialisation of cities, and smaller centres.

We examine these settings in detail in Chapters 3, 4, and 5, but, in summary, Cloke and Beaumont (2013) describe five types of response-spaces associated with postsecularity. First, there are places of *affective response* where the experience of obvious instances of socio-economic need induce a capacity to act within parts of the collective political and ethical conscience – places that cry out for 'something to be done about something', where a stark phenomenology of need prompts responses that deploy a willingness for rapprochement regardless of potentially divisive differences. Issues such as homelessness, anti-trafficking, refugee support, and food insecurity have worked in just this manner to create an affective capacity that prompts new intersections and crossovers between faith-motivated and other actors. Second, there are spaces of *resistance and subversion*, evidenced, for example, in research on services for homeless and unemployed people (Cloke *et al.* 2010; Williams 2015; Williams *et al.* 2012). Here, rapprochement across perceived religious/secular divides can occur both inside government funding schemes and beyond, but is organised in such a way as to undermine or deflect the neoliberal politics involved, for example, by providing an excess of care beyond stipulated limits, or by using charitable resources in contravention of government ideologies (for example, serving homeless people on the streets when government is seeking to sweep such people off the streets). Third, there are spaces of *voluntaristic and charitable cross-subsidy* in which services located in more marginal areas of the city are delivered using resources from other places. Although spaces of postsecularity will often involve voluntary work by ex-clients, they depend mainly on labour and finance drawn from the participation and support of people from other areas of the city – often from the more affluent suburbs where the socially excluded victims of austere neoliberal regimes rarely present themselves in person. Of course, this kind of reterritorialisation can easily be critiqued in terms of guilt-tripping and 'moral selving' (Allahyari 2000) of the affluent middle classes; a minor charitable gesture that both absolves personal responsibility for the exclusionary nature of society and at the same time reinforces the divides that proliferate that exclusion. We would caution against such easy dismissal of charitable voluntarism, not only because it underestimates the sacrificial and affective nature of some voluntaristic involvement, but also because it fails to notice the capacity for gradual but attractional in-commonness (Popke 2009) that can develop therein. Jack Caputo (2001) points to the 'anarchic effects produced by re-sacralising the settled secular order' which are capable of producing an 'anarchic chaosmos of odd brilliant disturbances', which he sees as 'gifts that spring up like magic in the midst of scrambled economies' (p. 291). Rather than automatically characterising charitable cross-subsidy in the city entirely as a light-weight sop to the middle class conscience, we might pause to reflect on the possibility that in amongst this apparent charity there is scope for in-common encounters in which alterity is attended to in ways that can make deep impacts on conscience, ethicality and political conviction, and in registers where caritas and agape find anarchic expression that can disturb the scrambled economies of marginalisation and exclusion through performances of receptive generosity. In these kinds of ways, the flow of economic, social, and sometimes spiritual capital into the spaces of care in marginalised areas of cities develop socio-spatial connections which unsettle not only perceived religious/secular boundaries but also geographical ones.

A fourth aspect of the reterritorialisation resulting from engagements with postsecularity involves the development of spaces of *ethical identity*. Here, we would point to the specific campaigning that aims to sponsor new ethical tropes such as City of Sanctuary and Fair Trade City (Amin 2006; Squire 2011; Darling 2010; Malpass *et al.* 2007). There is significant evidence to suggest that these campaign spaces reflect a range of religious and other interests, brought together to express values (for example, relating to hospitality, generosity, and responsibility) at the heart of which lie significant points of ethical convergence between theological, ideological, and humanitarian concern. In this case the conventional neoliberalised practices of marketing and branding places are being challenged or even usurped by a repositioning of cultural-political identity, drawing on rapprochements grounded in postsecularity. Similar spaces of ethical identity can also be recognised in more localised and ephemeral practices and events, and so a fifth strand of reterritorialisation is found in the myriad spaces of *reconciliation and tolerance* involving individuals and groups who are working across, or at least problematising, previous divides involving inter-religious, anti-religious, or anti-secular sentiment. Examples of such crossing of divides include anti-sectarian spaces (such as Co-exist in Glasgow) and peace-making spaces (such as the Bridges for Communities events in Bristol and Cardiff), although we also need to acknowledge here the possible negative outcomes that can occur when radical hospitality becomes *too* radical (see May 2018). At its root, this strand points to the embodied performances of identity – faith-motivated or otherwise – in which local lived spaces can come to represent the potential for new forms of tolerance and agreement in place of previous sectarian tendencies. Although more attention is usually given to the spaces of postsecularity that are organised by specific groups of activists in well-signalled initiatives, this smaller-scale and less prominent performative embodiment of postsecularity and rapprochement is equally if not more significant in its likely impact on the ordinary and everyday geographies of different spaces.

These five types of spaces of postsecularity are neither mutually exclusive nor exhaustive, but they provide initial evidence of the possibility of emergent hopeful geographies of in-common seeking after the common good. The ethics and politics of Romand Coles' postsecular caritas seem to be finding expression both in some recognisable social movements, and in some more modest sites of ethical activity. While it would be wrong to suggest any one driving force behind these initiatives, the increased activities of FBOs are a significant component in the development of different forms of rapprochement, and are certainly making their mark in city-spaces and beyond, both in putting into praxis theological and theographical (Sutherland 2017) ethics and politics, and in opening out meeting-places for wider partnerships with other individuals and organisations. These interventions are now part of what the contemporary city *is*. Given that the city represents a scale of religious identity and practice that best permits the organisation of faith-motivated action, both within and beyond faith boundaries, there is clear need to examine and assess the significance of this re-emergence and reformulation of the sacred within the wider development and spaces of urban communities; how new forms of re-engagement of faith and politics, or faith and ethics, are sponsoring particular virtues of in-commonness and common good within areas of governance, service delivery, and social protest. Moreover, despite scalar disadvantages, it is important not to ignore how such rapprochement is also taking place in more rural environments (see Jones and Heley 2016), where action, for example, on homelessness (Cloke *et al.* 2007), and food poverty (Williams *et al.* forthcoming) is beginning to gain pace. As well as reflecting how things are, the lens of postsecularity also offers insights into how the contemporary city (and its more rural counterparts) *could be*. Baker and Beaumont (2011b) suggest that spaces of postsecularity in the city serve as liminal spaces in which citizens are able to journey from the unshakeable certainties of particular worldviews with their extant comfort zones, to the unknown real and imagined spaces of rapprochement. So, for example, volunteers in spaces of care such as food banks will often find their initial political and/or religious assumptions questioned as they enter into relations of in-commonness both with hungry clients and with other volunteers with different assumptions (see Williams *et al.* 2016). In this way, partnerships of postsecularity could represent laboratories in which elements of self-interest and control are ceded to the greater aim of 'doing something about' social injustice and caring for others, and in the process, there are possibilities for transformations in ethical discourse and praxis. These ethical laboratories could be built on ideas of receptive generosity and rapprochement, which offer scope for reterritorialising socially divided spaces. However, there is a crucial third strand that underpins the idea and practice postsecularity – the potential for re-enchantment.

Assemblages of hopeful re-enchantment

In their account of the relationship between religion and contemporary activism in the Occupy movement, Cloke *et al.* (2016) argue that religious involvement helped to facilitate progressive crossover narratives that enabled both to discern the spiritual aspects of capitalism, and to galvanise prepolitical values of hope, faith, and love in the context of prefigurative projects of economic democracy and social liberation. In recognising capitalism as a regime of desire, emergent politics and ethics of postsecularity engage with the possibility of working towards a reformulation of prepolitical desire around notions of receptive generosity, respect for alterity, and reaching out with agape and caritas to neighbours and to enemies; that is, postsecularity works towards the reshaping of subjectivities by combatting neoliberal structuring of desire with new resonances of re-enchantment.

Weber's (1976) argument connecting capitalist modernism with processes of disenchantment is very familiar; in a disenchanted world, he argues, public life has been stripped of ultimate and sublime values as scientific rationalism and bureaucracy have replaced the magical, the mysterious, and the incalculable. Commentaries on the impacts of contemporary secularised neoliberalism have pointed to just such disenchantment in the current age. According to Blond (1998) – an author who identifies both as Christian and as 'Red Tory' – secular frameworks for advancing science and economics have been reproduced into the arenas of politics and ethics, risking a dangerous complicity with an ontology of violence that champions self-centred individualism and standardises the priority of force and counter-force. The result is a kind of hopeless vacuity in which the weakening of mysticism becomes characterised by endless self-serving acts of negation and denial, and self-seeking desire fed by the machinic power of commodity fetishism. Cloke and Beaumont (2013) regard this hopelessness as 'a weary acceptance in some quarters that how we live is circumscribed by the market-state's ability to shape how we govern ourselves', implying for many a 'broad disavowal of any possibility that social melancholia and desperation might be transformed or transfigured' (p. 39 – see also Milbank 2006). In a collective sense, the form and content of neoliberalism have placed boundaries on what pleasure can be and undermined more hegemonic regimes of desire. In such a context, Critchley (2012) and Ward (2009) describe the paralysis of empty nihilism in which the only discernible telos lies in the individual pursuit of pleasure. This individualisation of desire, and the associated erosion of notions of the community and the broader social, have resulted particularly in intensified class divisions often tightly constrained within social bubbles of sameness; living alongside people with similar circumstances and modes of dwelling, and not only reducing encounters with less privileged others but also actively feeding the neurosis of autonomous desire that often leads to stigmatising stereotypes of others-as-enemies (see Reinhard 2005). Re-enchantment, then, involves attention to both self-to-self and self-to-other pleasure and desire.

Weber further argues that the disenchanting work of capitalism is reinforced by its ties to particular forms of religion, and these ties have been increasingly and destructively imbricated in the era of neoliberal austerity. Connolly (2005 and 2008) describes how this Evangelical-Capitalist Resonance Machine in the United States represents an alliance between dogmatic 'cowboy capitalism' and right-wing evangelical Christianity that shores up and inspires the politics of existential resentment and hubris that have been largely responsible for much of the economic inequality, socio-cultural enmity, environmental degradation, and short-sighted parochialism of contemporary life. The capitalist axiomatic of prioritising private profit, granting unquestioned credence to market forces, and commodifying unjust labour formats is justified and stretched out by a form of judgemental and vengeful right-wing religion that too often espouses individual prosperity and extreme moral conservatism. The one resonates with and amplifies the other in a darkly reactionary form of fusion. In one sense, this crossover between conservative Christianity and conservative politics might be regarded as 'postsecular' in nature, but its complicity with a lack of interconnection between self-to-self and self-to-other desire contrasts strongly with our notion of 'postsecularity', that is, a condition of being that prioritises receptive generosity over exclusive ownership of theo-ethics.

The idea that society has travelled from an enchanted past through a disenchanted present to a possibly re-enchanted future is, of course, an oversimplistic model of social and cultural change that easily underestimates the ambivalences and overlaps that best of each of these supposed categories. However, the possibility of re-enchantment, including the return of magic, mystery, and irrationality, has been much discussed. As Lyons (2014) indicates, many such attempts to reimagine enchantment have in effect sought to revise the secularisation thesis, not by challenging the idea that modernity has inflicted fatal wounds on religion, but by contesting Weber's identification of modernity with a lack of enchantment. A stream of recent writing (see for example, Bennett 2011; Landy and Saler 2009; Levine 2008; Saler 2011) has suggested that modernity does not straightforwardly disenchant the world; rather it produces an entirely new array of secular strategies that yield often superior – if paradoxical – versions of enchantment. According to Levine (2008) and Bennett (2011) respectively, these secular enchantments inspire an excited affirmation of things in the world, and motivate ethical and political engagement in opposition to, or stretching beyond the capitalist machine. To some extent secular re-instatement of enchantment leans on the capacity of the postmodern to restore to the world what modernity had denuded it of; a reinvigoration of initiative and authorship of action, and the right to give meaning to and construe narratives counteracted a de-spiritualised and de-animated model of the world in which the capacity of the subject had gone missing.

A significant direction in the development of these geographies of secular re-enchantment has been the willingness to address the issue of a restored ontology of agency in both the human and non-human world. This return to an enchanted secular *Cosmopolitics* – or, in Blok and Farias' (2016, 5) terms, an 'ontological' or 'object-centred' politics – has its roots in the work of the Bruno Latour and Isabelle Stengers in the early years of this century, which is further developed by McFarlane (2011) and his ideas of urban assemblage. A fully-fledged ontological politics not only taps into the desire for a non-materialistic and more ethically grounded form of civic and political participation that we are suggesting lies at the centre of postsecularity. It also highlights the progressive and innovative potential of new assemblages of *Cosmopolitics* that often centre on everyday struggles of survival and dignity, especially for those on the margins.

Baker (2018a) has identified four significant aspects of these ontological politics. First, as suggested by Blok and Farias (2016), the human and non-human objects that co-construct our urban assemblages are not 'objects' in the standard sense of the word; rather they are relationally intended, and in shaping our 'shared, common public matters' they therefore come 'loaded with moral and political capacities' (p. 7). This radical co-presence helps to shape how urban realities are made and remade, and in turn generates a series of innate and virtual possibilities leading to a new surplus of knowledge and affect that transcend existing binaries and old signifiers. This *excess or surplus of meaning and affect* is the second significant aspect of the ontological politics of secular re-enchantment, and in turn leads to a third, given that the excess or surplus of meaning and affect impels us to want to make their inherent potential more *visible*. Blok and Farias are clear that this visualisation of the inherent and the potentially possible is not only a research imperative (2016, 5), but also a politico-ethical one that in turn generates *new politico-ethical subjectivities*, based on a common understanding of the singularity of moral intent and ontological depth that lies beneath the urban. Fourth, these emerging political-ethical subjectivities, involving as they do the decisive decentring of the human subject and previous knowledge about how material reality is 'produced', in turn create *new forms of political imagination and praxis.* 'It is important to stress,' say Blok and Farias (2016, 7), that 'this ontological multiplicity does not just point to the different furniture of human worlds, but to different ways of "being human", of assembling and enacting humanity.' This manifesto for a new political imagination it seems to us, could, and does, potentially 'cross over' (Cloke 2015) into religious, spiritual and non-religious sites of practical urban engagement; 'a politics of exploring and provisionally settling what does and does not belong to our common [urbanised] worlds' (Farias and Blok 2016, 7). Chapters 3, 4, 5, and 7 explore and analyse in greater detail the many dimensions of this new ontological politics of postsecularity, containing as they do these four elements of radical co-presence, an excess of new knowledge and affect, the emergence of new politico-ethical subjectivities, and new forms of political imagination and praxis.

However, alongside this secular turn to an ontologically-heavy politics based on the 'real' and 'moral' power of each object within any given urban assemblage, a significant segment of the envisioning of re-enchantment has embraced the values and critical capacities of left-leaning religion, arguing that a crisis of secular consciousness requires new spaces and subjectivities of spiritual disobedience in which hierarchies, dispositions, and deeply held beliefs can be reworked. In their evaluation of Occupy, Cloke *et al.* (2016) exemplify this kind of spiritual disobedience:

> We suggest that Occupy needs to be understood at least in part as a deeply spiritual and sacramental protest, not solely in its aims and objectives, but in its practices, its hospitality to otherness, and in its offer of direct experience of mutualism and radical democratic forms of organising. The solidarity practices within encampments offered a deeply spiritual counter-formation to the affective repercussions of capitalist liturgies (or discourses) that saturate our everyday lives. Counter-neoliberal liturgies that enforce an alternative spiritual and ethical worldview to the neoliberal entreaty to consume, behave and be comfortable can be a pragmatically meditative resource for producing a hopeful subjectivity, that operates beyond a symbolic understanding or attachment to the capitalist order, recognising its perversity, and more able to imagine and embody prefigurative possibilities for living.
>
> (p. 516)

These ideas about new prefigurative politics and ethics involving progressive elements of religion are firmly embedded in Connolly's (1999) answer to the Evangelical-Capitalist Resonance Machine. He articulates the need to imagine, narrate, develop, and experiment with a new movement of the democratic left that will be organised across religious, class, gender, ethnic, and generational lines, and will be infused with an ethos of radical pluralism. He contends that the distinction between secular public and religious private life needs to be reworked, and that traditional ideas about unity and solidarity need to be translated into drives to form new kinds of provisional assemblages comprised of multiple constituencies and creeds. In identifying this 'politics of becoming', Connolly, a self-identified non-theist, notes the importance of the Christian left to an alternative kind of resonance machine; he advocates 'movements back and forth between registers of subjectivity' (including religious and other registers) in order that 'each infiltrates into the others' (p. 148), recognising the potential for re-enchantment arising from this process. He draws both on a meliorism that prioritises collective reflective action, and on Deleuzian notions of immanence in a world that is always becoming. Taken together, moments of reflexive and coincidental postsecularity can emerge as key components of re-enchantment. Religious and secular dispositions can be made to resonate in ways that re-enchant a world stripped of its mystery, ineffability, and virtually impossible.

Connolly's new politics of becoming resonates with the other strands of postsecularity discussed above. Receptive generosity is a prerequisite for reflexive inter-subjectivity. A willingness and capacity for rapprochement prefigures any tactical pursuit of reflexive and coincidental hopefulness. At its core, however, the seeking after of assemblages of hopeful re-enchantment involves a remodelling of the subjectivities of desire away from those inflicted by capitalist hegemony, and towards those that are fed by a counter-cultural, and sometimes theological, ethic that confronts and secedes from neoliberal regimes of desire (prioritising wealth, self-interest, and self-pleasure) in order to cultivate an affective capacity for hopefulness and healing, hospitality and generosity, justice and equality. This process involves not only a reshaping of desire, but also experimentation with new affective rhythms and with new capacities to be affected by those rhythms. Such a remodelling will be connected with the cultivation of a pre-political ethicality relevant not only to material life, but also to an affective psycho-spiritual life; ethicality that reflects a protean movement of hopeful postsecularity that begins to take shape in and around impossibility as well as possibility, and indeterminacy as well as determinacy, as found in the geographies and temporalities being moved through and that contextualise becoming (Holloway 2011a). In one sense, following Rose (2017), re-enchantment may not be the best term to describe this new hopefulness of postsecularity, given that the 'entangling networks which constitute contemporary capitalism function as a system of technological re-enchantment' and constitute a 'secular reiteration of the kinds of structures of power and domination which characterised the enchanted universe of classical Christian thought' (p 243). As we discuss in Chapter 3, postsecularity perhaps sits more easily under Rose's banner of 'post-disenchantment', recognising a break from both classical Christian enchantment and from subsequent re-enchantments that reinforce the resonances between neoliberalism and the religious right. As our book progresses, we will examine evidence of contemporary post-disenchantment evident in lines of flight associated with the subjectivities (Chapter 3) and spaces (Chapter 4) of postsecularity.

1.4 Geographies of postsecularity

In what follows we examine these three interconnected modalities of postsecularity – receptive generosity, partnership, and re-enchantment – following a number of different trajectories. Chapter 2 presents a detailed exploration of the philosophical underpinnings of the concept. It traces key moments of debate concerning the role of religion in the public life of liberal democracies, beginning with the insistence (for example, by John Rawls and Robert Aldi) that religious narratives and ideas should be subjugated to and translated into secular equivalents so that the privilege of public secularity could be maintained. It then proceeds to the seminal, but much debated contribution of Jurgen Habermas; for some little different to the anti-religious defence of the secular that went before, but for others the protagonist for a more even-handed approach to religious and secular roles in public thought, and the 'godfather' of postsecularity. We discuss in detail his key ideas, especially his explanation of how crossover narratives emerge from the engagement of mutual tolerance across religious/secular boundaries, and of how mutual translation across these boundaries permits reasoning that can be accepted by religious citizens and secular citizens alike. Mutual translation, then, leads to complementary learning, which in turn facilitates a potential assimilation of ideas and a reflexive transformation of thinking which can lead to the acceptance of religious ideas into public policy. Rather than envisaging any wholesale acceptance of religious legitimation for public action, we understand postsecularity as a capacity to hold together the combined discourses and praxis of secular and religious citizens, enabling broad-based alliances to be built on a willingness to focus on ethical sympathies and practices, even if that means setting aside potential moral differences. For such alliances to achieve anything approaching rapprochement, secular and faith-based fundamentalisms need to be set aside, allowing primordial ethical and political currents to emerge in the pre-political realm that can subsequently flourish without being crushed or co-opted by neoliberal capitalism. We argue that the spaces and subjectivities of postsecularity may be most evident in the traces, flows, fragrances, and affective tolerances that are formed out of a mutual sense of theopoetics, and become part of Connolly's politics of becoming in which new energies and lines of flight emerge from the power of powerlessness, the possibility of impossibility and the translation of theo-ethics of peace, generosity, forgiveness, mercy, and hospitality into everyday praxis of care and justice for the other.

In Chapter 3 we examine the *subjectivities* of postsecularity as part of the wider conceptual recalibration of the notion of the postsecular. We begin by drawing on poststructural and non-representational approaches to subjectivity, affect and ethics to provide a definitional discussion of what is meant by subjectivities of postsecularity. We take postsecularity as a thirdspace where the blurred boundaries between religious and secular belief, practice, and identity can undergo reflexive engagement and produce new ethical and political subjectivities. By nature, religious and secular subjectivity has always been mutually co-constituted; yet, we contend that the intensity of new forms of enchantment, the deterritorialisation of propositional modes of religious and secular belief and practice, alongside postures towards receptive generosity, represents something that demands academic attention. We identify a series of tectonic shifts that have led to the erosion of religious and secular fundamentalisms and the bubbling up of subjectivities of postsecularity. We wish to draw attention to the constitution of secular and religious subjectivities under late-capitalist and neoliberal regimes of desire and suggest ways in which *existential ressentiment* and *spiritual ennui* might facilitate, as well as serve as a bulwark to, a willingness to enter into the space of postsecularity. The chapter then traces four possibilities presented by postsecularity in this existential and political-economic context: (i) as an opportunity in neoliberal austerity to retain practices of solidarity and amplify an ethics of in-commonness associated with new movements beyond religious and secular fundamentalism; (ii) as a mode of active resistance to neoliberal subjectification which has become marked by an ignorance and vindictiveness towards the 'other'; (iii) as an affective politics of hope that cuts through prevailing affective atmospheres of neoliberalism to solicit a 'hope of the hopeless' grounded in the non-foundational theopoetics of the impossible; and (iv) as a mode of post-disenchantment, where we explore the new ethics, values, and practices emanating from dissatisfaction with allures of religious and late-capitalist enchantment. We then explore how new ethical capacities and subjectivities of postsecularity emerge in practice. Here we draw on examples of the Pauluskerk, a faith-based harm reduction project in Rotterdam, The Netherlands; practices of solidarity and rapprochement in the Jungle camps in Calais, France; the work of the Third Sector in post-disaster Christchurch, New Zealand. We conclude the chapter by discussing how might our theorisation of postsecularity benefit from recent geographic work on the event.

Chapter 4 examines the variegated geographies of postsecularity, drawing on a series of empirical examples to illustrate how distinct spatial contexts afford different opportunities and barriers for subjectivities of postsecularity to emerge. The chapter begins by mapping a range of social and political spaces which have been incubators for subjectivities of postsecularity. This list is not intended to be mutually exclusive or exhaustive. Rather our purpose is to examine how postsecularity is differently produced, mobilised towards different ends, and generative of specific formulations of political and ethical subjectivity. To illustrate this, we compare two spheres of activity – neoliberalised spaces of welfare, and environmental activism – as a way of foregrounding the different discourses, practices, and spaces that curate different possibilities for partnership between religious and secular voices. We then offer a critical examination of how the three modalities of receptive generosity, rapprochement, and the reconfiguration of desire are manifest 'on the ground'. This is developed through two case-studies drawn from the arena of welfare and care in the UK: drug and alcohol treatment and recovery, and emergency food aid provision. In each arena emergent postsecularity is shown to be generative of, and shaped by, different configurations of the religious and the secular, opening out distinct political and ethical possibilities. Within this, however, we draw attention to the power entanglements and contradictions often embedded in rapprochement.

Chapter 5 examines postsecularity in terms of its progressive political utility. We begin by recognising that the centrality of ethical negotiation in postsecular politics is what distinguishes it from other political modalities. We argue that this ethical negotiation emerges from a common political and affective conjuncture that binds people across different ethico-political predilections and generates new modes of affective and political being which have been examined in recent research on activist practices such as community organising and hospitality. We use the example of participatory art practice as a way to illustrate that the postsecular transformations can occur in mundane and creative spaces as well as more obviously politically charged ones. We argue that spaces of care, quotidian rhythms, and creativity can be facilitated by politically minded activists, translating ethical proximity and affective commonality into politically charged action. This looks different in different postsecular modalities. Regarding re-enchantment, we give the example of religious activists grounding the speculative ontology of radical theorists such as Žižek (2000) and Badiou (2003) in order to question the legitimacy of hegemonic constructions of the Other. In the case of receptive generosity, we examine the use of 'techniques of self' (Connolly 1999; Foucault 2005) in religious communities whose communal structure empowers and practically supports heterogeneous forms of activism. Finally, apropos of rapprochement, we analyse the creation and development of crossover narratives in the Occupy movement, which served a pragmatic role in focussing action and subsequently evolved so as to generate novel subjectivities, partnerships, and political tactics. To conclude the chapter, we suggest – first – that postsecular analysis and tactics can help to forge new 'progressive coalitions' (Klein 2017) in politics by seizing back the psycho-spiritual terrain of politics from neoliberalism. Second, we suggest that by seizing back the terrain of psychology and spirituality from neoliberalism, new ontological approaches to politics can be embraced by progressive political movements that blend liberated desire, sense of place, and ethical sensitivity. Finally, we argue that these developments can generate new possibilities for participatory action research that blends ethics, politics, and spirituality.

In Chapter 6 we interrogate a central critique that associates postsecularity with a perceived Western and Euro-centrism, and the argument that it loses cultural and critical traction once it leaves that privileged locus. We echo calls for empirical sensitivity towards the spatial formations of postsecularity beyond the Christian-secular nexus of engagement. Equally, we warn against making normative assumptions about how postsecularity works out in practice and emphasise the danger of theorisation that is ungrounded in sociohistorical contingency. However, we argue that the global and hegemonic attempts in the nineteenth and twentieth centuries to impose a series of discrete models of Western secularism on colonial and now post-colonial cultures have left their imprint on the twenty-first century, creating a series of affective, spatial, and policy practices across non-Western social-spatial and historical contingencies. These provide a more nuanced analysis of our interpretation of postsecularity as receptive generosity, rapprochement/partnership, and re-enchantment.

The chapter identifies key aspects in the changing geographies of secularism, secularity, religion, and belief which produce highly localised and uneven capacities for postsecularity to emerge in different places. To highlight these contingencies, we trace our three-fold definition of postsecularity through existing literatures and key debates. First, we discuss gender identity and agency with regard to postsecular feminism and the identities of Muslim women in Turkey, and elsewhere. Here we note the entangled religious and secular identities and public performativities and highlight the possibilities for political conscientisation and new forms of rapprochement between secular and religious Muslim women. Second, we consider the potential rapprochement linked to changing perceptions in international development and humanitarian studies of the traditional relationship between the religious and the secular. Third, we suggest global variants of secularism are becoming increasingly fragmented and blurred through changing geographies of modernity, religion, and belief, and highlight the changing state-religion relationships in China, Russia, and India to raise questions about the localised and highly variegated possibility for postsecularity to emerge in situ. Last, we draw on perspectives of assemblage and actor network theory to expand our understanding of the ways that 'the religious' reproduces urban modernities, and vice versa. Through these illustrations we seek to offer a critical perspective that is attentive to the diverse and ambivalent terrain upon which the three modalities of receptive generosity, rapprochement, and hopeful enchantment might be curated.

We conclude by suggesting that the differentiated and lived out fields of the religious and the secular identified by Bourdieu, Casanova, and Taylor, so characteristic of previous understandings of modernity, are becoming increasingly blurred and hybridised across global settings. We end by drawing on the work of post-colonial literary theorist Manav Ratti who foregrounds emerging postcolonial literatures which are marked by new forms of hopeful ethics that seek new conceptions of secularism and religion. We go on to develop this theme of postsecularity as spaces and politics of hope in the final chapter.

Chapter 7 concludes this volume by analysing how the arguments we have built up throughout our narrative demonstrate how postsecularity is, first, affecting politics presently, and second, creating possibilities for future political and academic action. We address contemporary politics by reviewing Chapters 1 to 6, not to reiterate our arguments but to present the specific political salience of postsecularity in each chapter, focussing particularly on how its renegotiation of the secularism/religion interface fosters its political importance. Regarding future political and academic action, we present four arguments. First, we argue that the performativity of postsecularity can generate networked resonances which can impact the predominant political affects of late capitalism – fear, ressentiment, listlessness, and ennui – supplanting it with love, hope, and enchantment. Second, postsecularity opens up activism to an a/theistic spirituality that could reinvigorate the Left by reconnecting praxis with desire and steering new practices of generosity away from 'dark' postsecularity. Third, postsecularity can provide new maps for political analysis and action, generating a greater understanding of, and openness to, the fluidity of praxis. Finally, we provide a series of questions that postsecularity raises for human geography, creating new directions for research that interrogate the blending of religious and political concerns in critically appraised and ethically pluralised ways.

2 Genealogies

2.1 Postsecularity: complexity and confusion

In this chapter, we discuss in detail some of the principal theoretical and philosophical ideas that provide a foundation for our focus on postsecularity. An
emergent stream of multidisciplinary discussion over recent years about the
postsecular – much of it laced with gate-keeping polemic from the protectors
of various disciplinary faiths – has often led to confusion rather than clarity,
and to deeply conflicting appreciations of how ideas about the postsecular can
contribute to contemporary understandings of religion in society. For example,
Jurgen Habermas (2010) points to a rather optimistic vision of the postsecular
as a repository of the progressive democratic values that can emerge when
the sum of religious and secular reasoning becomes more than its constituent
parts:

> Among the modern societies, only those that are able to introduce into the
> secular domain the essential contents of their religious traditions which
> point beyond the merely human realm will also be able to rescue the sub-
> stance of the human.
>
> (p. 5)

By contrast, James Beckford (2012), offers a rather more downbeat assessment
of the idea of the postsecular, arguing that:

> The meanings attributed to the 'postsecular' are not only varied and partly
> incompatible with each other but also incapable of explaining the interpella-
> tion of public religions in Britain – and, possibly, elsewhere in the world.
> The concept of 'postsecular' trades on simplistic notions of the secular. It
> has a shortsighted view of history. It refuses to examine the legal and polit-
> ical forces at work in regulating what counts as 'religion' in public life.
> There is therefore a danger that talking about the postsecular will be like
> waving a magic wand over all the intricacies, contradictions, and problems
> of what counts as religion to reduce them to a single, bland category.
>
> (p. 17)

The postsecular as rescuer of mankind, or as illusory magic wand? These contradictory assessments can only be evaluated with a deeper understanding of how the concept and its terminologies carry specific meaning, and why different commentators choose to confirm their predispositional analyses or recognise the value of conceptual innovation in their evaluations of the phenomenon.

From the outset we need to emphasise that theorisation of the postsecular has stemmed from a diverse body of scholarship influenced by a wide range of theoretical and political frameworks and worked out in contrasting empirical contexts (see, for example, Barbieri 2014). So, our task in this chapter is to provide a clear and unambiguous journey through this diversity, in order to position our particular claims about postsecularity. At the most foundational level, our focus on postsecularity rests on a broad recognition of the convergence of the two supposedly incompatible trends of continued secularisation and a sustained (and perhaps increasingly significant) role for religion in politics and public affairs; circumstances which Elaine Graham (2013) characterises as placing scholars between a rock (of religious resurgence) and a hard place (of institutional decline and secularism). To recap, in Chapter 1 we discussed 'religion' in terms of conditions of being and systems of belonging that seek to interconnect human beings with the spiritual and the transcendental. Our discussion of 'secular' focussed on the political project that denies religion any privilege in the affairs of state. The idea of the postsecular, therefore, represents attempts to grasp the importance of how secularity and religion co-exist – or even work together – in what are perhaps new ways that relate to contemporary contexts of political-economy and society-culture. However, as with the long-standing discussions about the exact nature of the secular, various (and sometimes competing) strands of meaning have been developed about the postsecular, including reference to the interrelated concepts of postsecular theory, postsecularism and postsecularisation. As Olson *et al.* (2013) point out, each of these ideas has a distinct contribution to make: postsecular theory involves a rigorous critique of secularisation theories; postsecularism reflects a normative positioning with regard to the involvement of religion in public life; and postsecularisation indicates a re-emergence of religion in the public sphere. In other words, postsecular tendencies can be recognised in terms of conceptual ideas that extend existing debates about the secular, deep-seated ideologies about the public role of religion, and empirical accounts of the ebb and flow of religion in secular society. Clearly, in each case the notion of the postsecular rests on highly complex and contentious questions about what religion is, how it is assembled as a co-constitutive factor within secularity, how the changing nature of that assemblage should be recognised and assessed, and what credence should be allocated to the contribution and momentum of faith-based values and activities in this dynamic landscape. Perhaps most importantly, we need to question whether are there significant signs that the supposed boundaries between the secular and the religious are becoming blurred in particular contexts, and what kinds of possibilities – both hopeful and regressive – are emerging from the resultant postsecularity of agency and affect.

The response of some scholars to these kinds of debates is to question alto-gether the validity of the construct 'postsecular'. For others, the response to complexity has been to invent new terms that engage with, but at the same time tread carefully around, the existing stable of postsecular ideas. For example, della Dora (2016) uses the term 'infrasecular' to reflect the multi-layered co-habitation and competition between multiple forms of belief and non-belief. As indicated in Chapter 1, our particular emphasis in this book is on postsecularity which we define as a condition of being that is evident in the context-contingent bubbling up of ethical values and practices arising from amalgams of religious and secular determination to relate differently to alterity. As such, we are content to collaborate with the progression of ideas forming around the notion of the postsecular, but we want to specify particular time-space contexts in which we see the 'condition of postsecular being' emerge. So, while we are seeking to use postsecularity to shine a critical light on the normative assumptions of secular ideology, we are working at a much humbler scale than any wider claims about temporal or spatial sea-changes. To be clear, we do not regard postsecularity as indicating a new epoch that replaces the secular, or as an overarching descriptor for nations or city-spaces. Neither do we support any use of the idea of postsecu-larity as a generic predictive concept that raises expectations that all possible partnerships between secular and faith-based interests are somehow being trans-formed into a state of willing rapprochement. What we are convinced of, however, is that postsecularity serves as a potent heuristic device with which to discuss both the ways in which certain spaces and subjectivities become infused with a receptive generosity between religious and non-religious organisations and people, and the potential for progressive and re-enchanting partnerships. Such partnerships work with alternative (and sometimes even premodern) senses of desire, values, and affective hopefulness to those of the neoliberal norm, in order to bring about ethical practices that transcend the sum of their parts, and in so doing suggest new ways to relate differently to alterity and to become active in support of others.

In the remainder of this chapter, we examine a number of different but over-lapping conceptual issues that help to underpin the notion of postsecularity. We do so as a critical engagement with foundational ideas rather than as a defensive posturing against the criticisms of other commentators; nevertheless, it is useful at this point to rehearse some of the principal reservations that have been used to question the validity and utility of postsecular ideas, so that these arguments can be woven into the critical reflections that follow. There are five strands of opposition that contribute in particular to this warp and weft of postsecular critique:

i The concept of 'postsecular' trades on simplistic notions of the secular (see Beckford 2012). In particular, scholars of religion are often thought to assume that secularisation is marginal to the understanding of contemporary religion, and it might therefore follow that using ideas about the postsecular acts as a corrective to allegedly defective frameworks of thought about the secular. By contrast, Wilford (2010), for example, argues that secularisation theory continues to offer geographers of religion a powerful theoretical framework for analysing contemporary religion. In Section 2.2, we examine the secularisation thesis and its recent revisions, noting both the warning from Olson *et al.* (2012) that in revising this thesis the postsecular risks duplicating rigid categories and assumptions, and Mahmood's (2009) argument that to rearticulate religion in a way that is commensurate with contemporary sensibilities and governance will also require a rethinking of the secular and its truth claims.

ii The postsecular reflects a normative desire to bring religion and spirituality back into the centre of social concerns (see Danforth 2010). At its most basic, this anxiety points to the use of the postsecular as a construction formed by a section of the intelligentsia (Martin 2011) to promote unwanted religious priorities into secular publics. In Section 2.3, we review arguments for and against the restraint of religion in the public sphere, and focus in on the specific ideas of Jurgen Habermas – the godfather of the postsecular – who, having previously rejected the idea of public religion, presents a series of technologies that can be used to understand some of the basics of the receptively generous rapprochement that contribute to postsecularity. There follows (in Sections 2.4 and 2.5) a review of how the postsecular has been variously enrolled into conservative and individualist political theologies as well as those directed more towards goals of the public and common good (Bretherton 2010a). However, our approach to postsecularity is not a guise to establish normative conceptions of the 'common good' or valorise 'good religion' and denigrate the 'bad', or to contribute to the double disadvantage (Modood 2005) experienced by minority religious groups. Indeed, Chapters 3, 4, and 5 demonstrate the importance of examining the partialities, deliberations, and tensions involved in 'actually existing' spaces and subjectivities of postsecularity, while simultaneously foregrounding the emergent ethical and political sensibilities that embody the capacities for rapprochement, receptive generosity, and hopeful re-enchantment.

iii The postsecular is not new; its empirical and theoretical concerns are already being addressed using existing conceptual vocabularies (see Kong 2010; Ley 2010; Wilford 2010). Any construction of sacred/secular relations as proceeding through distinct presecular, secular, and postsecular stages is unlikely to be helpful or accurate. Yet the complex nature of postsecularity inevitably sponsors different interpretations of the degree to which contextual time-space circumstances differ from each other, and longserving commentators can understandably be predisposed to argue against the shock of the new, preferring explanations of long-term intermixing of the sacred and secular. Cloke (2011a) on the other hand, emphasises how postsecularity has appeared to flourish in the socio-political landscapes of neoliberal austerity, thus opening out a more emergent understanding of postsecularity as (at least in part) a potential site of resistance to the stringencies and privatised excesses of contemporary politics (see also Baker

2016; Cloke *et al.* 2011b; Williams 2015). In Section 2.6 we discuss the importance of these kinds of context, both to an empirical understanding of postsecularity in present times, and to the kinds of 'messy middle' theorisation that these times are currently suggesting in geography (see May and Cloke 2014). Nevertheless, the potential criticism remains that these events can be, and are already being addressed using existing conceptual vocabularies, thus rendering the idea of postsecularity redundant. Our response here is that we use postsecularity as a heuristic device that provides the technologies, tools, and vocabularies with which to make sense of the ambiguities and relationalities between the religious and the non-religious in political and ethical life that are not currently best served through conventional grammars of analysis.

iv   The postsecular is overdependent on religion as the key to understanding wider experiences and cultures of spirituality (Bartolini *et al.* 2017). In many ways, modernity is becoming increasingly enchanted, magical, occult, and spiritual, such that expressions of 'spirit' and 'superstition' are already folded through contemporary social life – both within and beyond recognisably religious containers. In Chapter 3 we explore how the potential re-enchantment that accompanies subjectivities and spaces of postsecularity is associated with mundane embodied practices and experiences as much as organised religious activity. A restored ontology of agency, including non-human agency, suggests the necessity to understand more deeply how realities are made and remade; how the human and non-human assemblages that make our world produce excesses of meaning and affect – including those indicating postsecularity – that help to create atmospheres, becomings, and potentials that contribute to new politico-ethical subjectivities (see Baker 2018b).

v   The postsecular represents colonialism by other means (see Bugyis 2015; Mufti 2013). Some accounts of the changing visibility of religion seem over-reliant on the growth and increasingly politicised role of Islam in order to make a case for a global-scale resurgence of religion. The criticism can therefore be levelled at postsecular ideas in general, and Habermasian postsecularism specifically, that the gesture of inclusion of Islamic religions into western democratic process represents a form of imperialism which is premised on the logic of colonial appropriation of the religious 'other' to shore up an 'increasingly insecure, liberal nation-state and defend it against those who might seek to take advantage of its vulnerability' (Bugyis 2015, 25). Such an instrumental incorporation offers a highly misleading view of contemporary Islam 'as an expression of religious consciousness directed against the inroads of secularism', thereby ignoring the diversity of lived forms of Islamic religiosity in a way that 'close[s] off prematurely the possibility of a materialist and historical understanding of the present in the Islamic world and a critical engagement with it' (Mufti 2013, 11). Read in this way, the postsecular risks cultural assimilation or 'dissolution of the entire social space of the other' (Mufti 2013, 17) to adhere to the supposed core values of western liberal subjectivity. Arguably such a move fails to constitute a truly pluralistic democratic politics by pre-empting contestation and serves mainly to ensure the legitimation of the juridical power of the modern state. These critiques chime with broader criticisms that theoretical accounts of postsecularism predominantly rest on Western European assumptions of what constitutes 'religion' and the 'secular' – assumptions which have long been criticised for upholding a religious-secular distinction that conceals explicit ideological values (for instance, concerning public versus private faith), that privilege Christianity at the expense of other religious expressions such as Islam (see Asad 2003). We acknowledge that to date empirical studies of the postsecular have relied heavily on evidence relating to Christian religion, and indeed there is an associated risk that the postsecular can be that this gesture of inclusion of Islamic religions into western theory represents a form of imperialism – a kind of colonialism by other means. It is indeed the case that to date empirical studies of the postsecular have relied heavily on evidence relating to Christian religion, and there is an associated risk that the postsecular can be narrated as a kind of one-way conveyor-belt of Christian religious freight into the secular domain. Such a picture, however, ignores not only the complex circumstances of how other major religions inflect different ideas, practices and performances into particular contexts, and the ways in which religion itself can be and has been secularised. The research that underpins our account of postsecularity-in-practice in Chapters 3, 4, and 5 does indeed focus mainly on the contribution of Christian religion to emergent postsecularity in the UK, and we emphasise the context-specific nature of this research. However, in Chapter 6 we present an account of postsecularity in other international contexts, and here we seek to recognise and mitigate the risk of colonialism by stealth.

2.2 Revising the secularisation thesis

The demise of religion has been foretold ever since social philosophy began to grapple with the likely impacts of modernity. For example, the seminal thinkers of the nineteenth century, including Durkheim, Weber, Marx, and Freud, each contended that the advent of industrial society would be accompanied by a gradual fading of the importance of religion, ending with its erasure as a significant factor in the shaping and experience of social life (see Aldridge 2000; Bruce 1992). As societies industrialise, it was argued, religious ritual and observance will gradually erode, and public opinion will become increasingly impervious both to the appeal of religious organisations, and to the attempts by their leaders and remaining congregants to retain an influential public voice. From this philosophical platform emerged a conventional wisdom that declared the death of religion and placed secularisation (alongside urbanisation, bureaucratisation, and rationalisation) as a powerful force in the transformation of agrarianism into modernised industrial society. The death of religion thus became an integral part of social science inquiry throughout much of the twentieth century, and the thesis of secularisation became ingrained into the mantra of secularist academic research, embedding, according to Casanova (1994), three key assumptions into studies of relations between the secular and the sacred. First, the overall belief in and practice of religion would decline and become decreasingly influential. Berger (1969) attributed the probability of a gradual disappearance of religious adherence in modern times to a loss of plausibility – and therefore confidence – in the supernatural truth claims inherent in established religious belief. The result, he argued, would be a more sceptical and more secularised populace, and an accompanying secularising shift in the focus of remaining religious gatherings from metaphysical belief to more worldly needs. Accordingly, a second assumption was that religious practice and belief would shrink into private corners and lose their traction in the public sphere. In one sense, this privatisation was thought to be prompted by the increasing prestige of science and the rise in cultural relativism which served to loosen the loyalties given to religious institutions and their public posturing; accordingly, faith would be associated with the private realm rather than the more public declaration of institutional creeds. Another perspective (see Luckmann 1967), however, suggested that religion would at least in part change its location, away from publically recognisable containers and into the wider and more private spiritualities of psychotherapy, esotericism, and the like. In both cases, a third trend – the exclusion of religion from public processes of state and social formation – would be a natural move of modernist independence. Of course, these three sets of assumptions are founded on the highly questionable idea that society before modernity had a highly sacralised past. Martin (1969), for one, sees this idea as fictional and therefore as undermining the apparent simplicity of the secularisation thesis.

Embracing these assumptions and predictions, the secularisation thesis became a basic narrative of modernist self-comprehension (McLennan 2007; 2011). However, even the most passionate advocates of the thesis have accepted that the remarkable resilience and persistence of religion have necessitated a rethink. Casanova (1994, 11) reverses his previous allegiance by asking 'who still believes in the myth of secularisation?' and by recognising that religion is becoming deprivatised through its emergence as a public issue in some societies. Berger (1999) also declares a change of heart, choosing to emphasise the desecularisation of a world, much of which is as religious as ever. In the starkest terms of re-orientating critique, Stark and Finke (2000) declare the need to bury the secularisation thesis in the graveyard of misrepresentations and failed theories. The need for these kinds of revisions is in part associated with the multifaceted complexity of the thesis. Wilford (2010) argues for a distinction between secularisation (a social process concomitant with modernity), secularism (a set of beliefs and practices with their own history and politics) and secularity (the generalised sociopolitical compartmentalisation of religion). The seemingly simple original secularisation thesis has thereby become recognised as a much more complex and multidimensional amalgam of dimensions that frequently work independently (Davie 2007); it follows that the supposed 'death of God', the supposed disappearance of religion, and the supposed disengagement of religion from public spheres of influence, are assumed outcomes that connect with geographically uneven processes at rather different societal, organisational, and personal scales. The rather Eurocentric assumptions (see Berger *et al.* 2008) of the original thesis compound the likelihood of its rebuttal in terms of empirical evidence.

So, it is in the context of this complexity that a range of different commentators have suggested evidence that contradicts, counteracts, or makes more complex the central tenets of secularisation. Eder (2002 and 2006), for example, notes that the supposedly secular societies of Europe are in different ways becoming increasingly subject to religious discourse. Secularisation, he argues, has not caused the disappearance *per se* of religion, but rather its temporary disappearance from public view – resulting in the public religious voice being 'hushed up' due to its relegation from public societal debates and its occupation of private spaces. However, in the USA, religion has retained a more significant place in the public sphere, representing a block of political economic interests which has been sufficiently powerful to retain a place in public debate in amongst continuing trends towards secularisation. Moreover, Eder observes that religion in Europe has found its voice again, more recently re-entering the public sphere with confident, if multifaceted contributions, to public affairs. Likewise, Habermas (2002, 2005, 2006a, 2006b, 2008, 2010) argues for an 'impression' of religious resurgence based on three empirical trends. First, he explores the advance of orthodox religious groups within established religious organisations, including Hinduism and Buddhism as well as the main monotheistic religions. Second, he describes how fundamentalist adherence to literal interpretations of scriptures and rigid moral values lies at the heart of the most rapidly increasing religious movements around the globe. Third, he is exercised by the potential for innate violence over religious causes illustrated by practices and outcomes of terrorism committed in the name of Islam. In these trends and events, Habermas recognises how public consciousness is changing as an adjustment to the continued existence of religious communities in a supposedly secularised setting. He emphasises how European citizens are very well aware of the intrusive activities of fundamentalist religious movements in their sphere of influence and are often unwilling to accept secularist dogma about the disappearance of religion. As a result, he argues that religion is gaining influence in the public sphere at both national and local levels both as a community of interpretation – contributing to public opinion in moral and ethical issues – and as a community of service and care, carrying out welfare tasks both within and beyond formal systems of governance. He also sees religion as changing public consciousness in the area of immigration and asylum-seeking, raising issues of how immigrant cultures can be integrated into postcolonial societies, and how different religious communities can achieve tolerant co-existence in city spaces.

Both Eder and Habermas view these trends as evidence of postsecular activities, and (as we explore in Section 2.3) Habermas (2005) in particular proceeds to identify basic normative frameworks for how to participate in what he sees as a postsecular society and envisages a role for religious communities in promoting a postsecular understanding of society. So, despite evidence in countries like the UK that formal religious adherence and allegiance are declining, the possibility is raised here that a change of public consciousness has prevented the disappearance of religion. In part, this conundrum of falling numerical presence but rising influence is explained both by what Davie (1994) terms as a continuation of religious influence through believing without belonging, and by other forms of spirituality persisting outside of institutionalised religion (see Section 2.6). In addition, however, such influence may reflect something of what Graham (2013, 48) refers to as 'the irreducibility and transcendence of religious principles' which can point to 'a depth of moral reasoning unavailable to secular understandings'. What is clear, however, is that secularisation did not 'unmake' religion; indeed, each has been co-constitutive of the other over the changing canvasses of space and time. In the same way, if more recent changes in public consciousness towards religion suggest evidence for postsecularity, then that postsecularity should not be viewed as in any way 'unmaking' the secular. Again, the one shapes the other. Casanova's (2013) revision of the secularisation thesis is based on an acceptance of a current obsession with religion and reflects the co-constitutive forces that assemble aspects of the secular and the religious socio-economic trends such as globalisation, migration, and religious pluralism. In so doing he concedes the possibility of postsecularity in these assemblages and concludes that Habermas is 'reading the signs of the times and interpreting the zeitgeist with prescient accuracy' (p. 45).

Such re-interpretation of the secularisation thesis remains open to different interpretation, not least in terms of the kinds of geographies that are envisaged in dealing with the persistence of religion. Here it is useful to compare the geographical constructions presented by Wilford (2010) and Cloke and Beaumont (2013). Wilford's analysis recognises space for religion within secular geographies but models these spaces as archipelagos within the surrounding seas of modernity. Arguing that modernity separates the social into distinct spheres, in this account, religion is seen to persist at the scale of personal, family, or community belief, and therefore the enchantment that religion offers to the modern world can be found in these distinct island-spaces of the sacred. The form and function of religion is to exert its own authority and influence in its own social sphere. Cloke and Beaumont, by contrast, focus on the performance of religion as a resistance to neoliberal austerity, by directly engaging in care, welfare, and justice, and by supporting and enabling public debate on these kinds of issues. They find religion involved in cross-over networks and discourses in which the secular and the religious sometimes find common ground for re-enchanting rapprochement based on receptive generosity. The first picture is one which continues to prioritise ideas of secularisation, within which religion continues to persist in its own spheres alongside the secular but cut off from it by tides of modernity. The second envisages the bubbling up of subjectivities and spaces of postsecularity within secular society, seeking to build networks of common interest and common good. Each model offers ground for the continuing significance of geographies of religion, but only the latter accepts the possibility of postsecularity within the secular.

2.3 Habermas and the public technologies of postsecularity

The secularisation thesis in many ways sits comfortably with demands of exclusionist philosophers that religious reason be restrained in its influence on political decision-making in the public sphere. For example, Rawls (1997 and 2005) argues that understandable 'public reason' is imperative as an appropriate basis for liberal democracy; in these terms excessively religious or philosophical doctrines need to be excluded from political discussion on the grounds that non-adherents of these doctrines will find them inaccessible, incoherent, and inappropriate as forms of reasoning for particular practices and outcomes. Using the biblical parable of the Good Samaritan as an example, Rawls insists that there is little inherent value in the religious morality of the tale, and that it is only when overlap is apparent between that morality and legitimate liberal democratic values that it can be transformed into acceptable public reason. The implication is clear – religious discourse and values may reflect existing public reason, but they should not be used as founding concepts for such reason. Audi (1989 and 1997) also advocates restraint on religion in the formation of public reason; he accepts that some religious people will be motivated by their faith and belief but insists that core reasoning in the public sphere must be secular in nature. In both cases, exclusionist philosophy demands a common secular arena for the reasoning and practice of political conduct. Any underlying codes of morality that cannot be accessed and understood by all citizens represent a threat to the values of freedom and equality that underpin liberal democracy. In particular, religious discourses and codes are regarded as inherently divisive and as promoting the kinds of conflict and instability that can lead to the disintegration of the political community (see Ackerman 1994).

It is clear that this philosophical detection of secular public reason can be accused of dealing in questionable ideal types about how political discussion and conduct works in reality (see Stout 2004; Wolterstorff 2007), and in effect seems to exclude citizens from practising their religious faith as a motivational force in public life (McConnell 2007). However, in his early career Habermas was an equally strong advocate for secular reason in the public sphere as an integral component of modernity. As Graham (2013, 45) notes:

> In earlier years, his Marxist convictions steered him towards a broad sympathy with a classic Rawlsian position which required the creation of a non-confessional public space in order to ensure the most equitable conditions for the articulation of a rich and non-partisan discourse of citizenship and communicative democracy.

Later in his career, the turn of the century saw Habermas apparently transform from poacher to gamekeeper in his invocation of a postsecular incorporation of religious values into a renewed lexicon of virtue in the public sphere.

Habermas was not the first to invoke the term 'postsecular' – Beckford's (2012) archaeology of the term traces it back at least to Greeley's (1966) account of how small, neo-Gemeinschaft communities were able to flourish within a secularising and rationalising Catholic church. However, he has been commonly recognised (see, for example, Dillon 2010) as giving an inescapable momentum to the idea of the postsecular as a significant shift in consciousness about the role of religion:

> I use the expression 'post-secular' as a sociological description of a shift in consciousness in largely secularised or 'unchurched' societies that by now have come to terms with the continued existence of religious communities, and with the influence of religious voices both in the national public sphere and on the global political stage.
>
> (Habermas 2013, 348)

The claim here is not of an end to the importance of secular reason or of a return to pre-secular religious influence. Rather, the impression of a worldwide resurgence of religion is used to argue that religious reasoning offers a powerful and irreducible source of meaning and identity and should therefore be included in wider arenas of public communication. By the same token, Habermas (2010) laments the melancholy of secularist modernism, defining 'what is missing' as an appreciation of the potential for transcendental and/or metaphysical roots to contribute to ideas of justice and human dignity. In short, his advocacy for the postsecular appears to endorse a significant role for religion in public reasoning for two main reasons (see Calhoun 2011). First, he questions whether 'progress' can adequately be conceptualised in terms of secular exclusivity. Second, he queries whether it is possible to differentiate clearly between secular public reason and discourses of faith in the public sphere.

Habermas' postsecular contentions have been thoroughly critiqued on a number of grounds (see, for example, Cooke 2006; Eberle 2002; Lafont 2007; O'Neill 2000). At their most fundamental, these critiques point to the implausible nature of postsecular thinking (Joas 2008), especially in terms of Habermas' apparent reluctance to define the scope and extent of 'religion', and his lack of discussion about how ideas of 'public' and 'state' interconnect (Beckford 2012). In other words, is Habermas simply appealing to like-minded liberal religious and otherwise spiritual people for whom (in contrast to the 'fundamentalists' he frequently identifies) translation of faith-based motivation and discourse into secular spheres is relatively unproblematic? And is he conveying a rather loose sense of the public sphere in which the more complex legal and legislative arenas (with their attendant regulation of religious discourse and language) are conveniently sidestepped? Changes in these contextual and definitional parameters present very different scenarios for potential postsecularity, suggesting to some that Habermas has in mind a rather narrow revision of what had become an overconfident secularisation thesis, rather than any wider acceptance of a new role for religion in radically changing times (Harrington 2007). Indeed, Gordon

(2011) argues for a strong continuity between Habermas' earlier focus on the secular, and his later reference to the postsecular, the main difference between them being a transition from a Rawlsian contractual model of democratic public space to an appreciation that human beings in public spaces carry with them values that are irreducible to the secular science of communicative reason. Little wonder, then, that Mendieta (2010) applauds the seemingly inclusivist innovation suggested by Habermas that all voices – including religious ones – should be regarded as legitimate in pluralist debate, while at the same time, Taylor (2011) bemoans the seemingly exclusivist tendency in Habermas' work to make few concessions to religion amongst the continuing pre-eminence attributed to secularist reason. Essentially both claims are available from Habermas' treatment of the postsecular, the direction of travel largely depending on the normative standpoint of the critique, and on the empirical circumstances that are used to imagine and play out what the postsecular looks like on the ground.

It is notable that the hard-hitting critiques of Habermas' work from within sociologies of religion are contrasted by what are often more sympathetic readings from other social sciences and the humanities, notably international relations (Mavelli and Petito 2012; Wilson E 2014), theology (Graham 2018), and English literature (Kaufmann 2009; Ratti 2013) that draw on and critically develop Habermas' conception of the postsecular. In human geography, the argument about how to ground Habermas' ideas empirically has been crucial, and there are significant examples of how these ideas have been used as an endorsement of pluralism in the public sphere (see, for example, Gokanksel and Secor's 2015 analysis of Islamic geographies in this respect). Notably, a stream of research investigating the increasingly significant role of faith-based organisations (see for example, Beaumont and Baker 2011; Beaumont and Cloke 2012; Cloke *et al.* 2013a; 2013b; Cloke *et al.* 2014; Williams, 2015; Williams *et al.* 2016) has charted how the arenas of care, welfare, and justice have proved to be fertile ground for faith-motivated groups and individuals to forsake privatised forms of religion and to engage in more public advocacy and action on behalf of socially and economically marginalised people. Moreover, researchers have noted an openness for some of these groups to work with secular and religious others in pursuit of goals which attend to the shrinkage of state welfare and the ravages of neoliberal austerity. In these contexts, readings of Habermas' philosophical constructs of the postsecular have gained significant purchase, not as an indication of the theoretical importance of religious discourse *per se*, but as a series of technologies of receptive generosity that chime with the on-the-ground practices of social activists. It is these technologies of postsecularity that are of core interest to the thesis being advanced in this book, and that act as a foundation both for the kinds of political and civic responses that we map out in Chapters 3, 4, and 5, and for the call to action that we advocate in Chapter 7. So before moving on we want to detail the key ideas that inform the arguments that unfold in later chapters.

First, it is important to note that Habermas argues that religion is gaining influence both as a community of interpretation – contributing to public opinion on moral and ethical issues – and as a community of service and care; in both ways public consciousness is being changed, and the technologies of how participation in spaces and subjectivities of postsecularity are being worked out. He asks:

> How should we see ourselves as members of a post-secular society and what must we reciprocally expect from one another in order to ensure that in firmly entrenched nation states, social relations remain civil despite a growth of a plurality of cultures and different world-views?
>
> (Habermas 2008, 7)

Habermas' (2010) response lies in a learning process in which secular and religious mentalities can be reflexively transformed, rather than being maintained in dominant and subaltern positions respectively, and in which conditions are presupposed onto both the secular and the religious. In both cases, he argues against fundamentalist positioning. Secularist polemic of negative response to religious doctrine and action is seen as incompatible with shared citizenship and respect for cultural difference that lies at the heart of the postsecular balance. Secular citizens need to be able to discover, even in religious utterances, semantic meanings and personal intuitions that cross over into their discourses. Religious discourse equally needs to respect the authority of 'natural reason' as the fallible outcome of science and of egalitarian law and morality. This latter insistence has been interpreted to suggest that 'Habermas actually makes very few concessions to religion' (Beckford 2012), and indeed Habermas does admit that the divide between secular knowledge and revealed religious knowledge cannot be bridged. Indeed, it can be argued that his incorporation of religious voices is a form of instrumental protection of a democratic public sphere facing the challenges of both right-leaning secularist attitudes and the political marginalisation of Islam in European countries. Nevertheless, moving beyond this instrumental argument, we want to emphasise that not only does Habermas establish a framework of mutual tolerance as the foundation for postsecular rapprochement, but he also argues for the possibility of distinct crossover narratives between the religious and the secular – where different knowledges may not be bridged, mutually translating narratives can be deployed on which to found this rapprochement.

Second, in *An Awareness of What is Missing* (2010), Habermas explores the discursive technologies of these crossover narratives, suggesting that democratic strength rests on broad moral stances that arise from prepolitical sources such as from religious life. Rather than normative guidelines, these sources are significant motivational forces for political and ethical practices. If religious and secular utterances are mixed together in the prepolitical soup, they might also be liable to a process of 'mutual translation' (Reder and Schmidt 2010, 7). As Habermas (2010, 16–17) puts it:

> The philosophically enlightened self-understanding of modernity stands in a peculiar dialectical relationship to the theological self-understanding of the

major world religions, which intrude into this modernity as the most
awkward element from its past ... modern reason will learn to understand
itself only when it clarifies its relation to a contemporary religious consciousness which has become reflexive by grasping the shared origin of the
two complementary intellectual formations...

Habermas suggests here that the role of religion in the contemporary liberal state
cannot be confined to legalistic considerations of freedom of expression but
requires a legitimation of convictions that can be accepted by nonreligious and
religious citizens alike. In order to replace 'what is missing', therefore, postsecularity requires a technology of complementary learning in which the secular and
the religious involve one another.

Critics (for example, Barbato and Kratochwil 2008; Jedan 2010; Malik 2007)
worry that Habermas' analysis of what is missing is primarily concerned with
'the health of public politics and morality' (Beckford 2012, 10) rather than about
any respect for the integral value of religion in public life. In one sense, this turn
towards religion is being shaped by high profile politics of multiculturalism and
religious diversity which demand a more open recognition of religious elements
of political difference. Alternatively, Habermas' attraction to what religion can
offer has been seen as limited to that of a vague repository of moral values. It is
certainly clear that his desire for complementary learning is embedded within a
concern over the takeover by markets of an increasing slate of regulatory functions (Habermas and Ratzinger 2006) which he sees as leading to a depoliticis-
ing of citizen action and a dwindling hope that creative political forces will
emerge simply from secular critical reasoning. However, Habermas is clear in
his call for a 'conversion of reason by reason' (p. 40), as reason reflects on its
origins, discovers alternative directions, undergoes an exercise of philosophical
repentance and transcends itself into an openness towards other frameworks of
reason, including religious tradition. In particular, this transcending openness
involves the assimilation by philosophy of ideas from Christian religion:

> This work of assimilation has left its mark in normative conceptual clusters
> with a heavy weight of meaning, such as responsibility, autonomy, and jus-
> tification; or history and remembering, new beginning, innovation, and
> return; or emancipation and fulfilment; or expropriation, internalisation, and
> embodiment, individuality and fellowship. Philosophy has indeed trans-
> formed the original meaning of these terms, but without emptying them
> through a process of deflation and exhaustion.

(pp. 44–45)

It is this process of assimilation, then, that seems to offer a basis for rapprochement, whether longstanding or in new form. Assimilation can lead to different
outcomes: it could lead to a dilution and deradicalisation of religious concepts as
they assimilate with the process of secularisation (this was the underlying fear
that has in the past fuelled opposition by evangelical Christians to the so-called

social gospel – see Cloke *et al.* 2012); or it could re-connect religious concepts to their societal roots as part of a postsecular technology in which crossover narratives form the basis of new partnerships across the religious/nonreligious divide. We want to argue that these crossover narratives are perhaps more significant than some critics would credit as devices of assimilation and reflexive transformation on which emerging rapprochement might be built upon. Although rather rationalist and cognitive in nature and requiring the role of embodiment as a site and source of resistance (Mavelli 2012), the technologies of crossover and reflexive transformation offer crucial clues to the potential to be realised through the subjectivities and spaces of postsecularity.

If the acid test of Habermas' ideas about assimilation and reflexive translation is whether religious legitimation for policy and action is overtly accepted in non-religious arenas (see Wilford 2010), then the test is really only passed in the Western world in terms of the influence of the religious right in the USA (Berger *et al.* 2008; Dionne 2008). If, however, assimilation and reflexive transformation are open to discernment within crossover narratives relating to ethical issues, for example, as seen in campaigns such as those relating to fair trade, anti-trafficking, and fair wages which have held together the combined discourses and praxis of nonreligious and religious citizens, then the ideas may have more purchase in understanding emerging partnerships in the city. Accepting criticisms about the limited geographical reach and potential ethnocentricity of the concept, it remains possible that in the context of Western Europe Habermas' ideas about postsecularity and its attendant reflexive and discursive technologies can inform our understanding of the conditions in which crossover mutualities between religious and secular discourses can be enabled. This requires us to investigate the possibility that crossover narratives are occurring in particular arenas of discourse and praxis, and that they are permitting broad-based alliances to develop around a willingness to focus on ethical sympathies and actions even if that means setting aside other moral differences. Here we might expect to find some evidence, then, of a drawing back from different forms of secular and religious fundamentalism in order to forge alliances and partnerships on key ethical and political issues.

2.4 Conservative political theologies of postsecularity: the example of radical orthodoxy

The technologies advanced by Habermas offer, at least in theory, significant scope for embodying and practising the relations of postsecularity between religious and secular parties. However, the propensity for different groups to participate in such relations is governed by a number of factors, not least the interplay between religious and political normativities, and their impact on the capacity to find common ground with different others. Thus, the idea of the postsecular can be evoked in a range of circumstances, each reflecting very specific normative foundations and being subject to particular forms of political enactment. Graham and Lowe (2009) discuss how a balance can be struck between the desire to speak coherently and authoritatively from a faith-based positionality, and the necessity in a pluralist context to engage in discourses and politics of inclusivity. They note how the 'discipleship' of post-liberal religious interests involves a normative foundation of key beliefs, whereas (often more liberal) religious 'citizenship' involves a more apologetic task of critical theological reflection in order to establish the relevance and credibility of religion in public arenas. Examples of these political theologies of discipleship and citizenship can be seen both in debates within Radical Orthodoxy about the suitability of theocentric forms of political action, and in the crossover narratives between contemporary political philosophy and poststructural theology that lead to rather different uses of religious iconography, discourse, and participatory action. These debates provide a useful illustration of how different versions of political theology variously exhibit different levels of receptive generosity, leading in turn to contrastingly regressive and progressively hopeful forms of postsecularity.

Radical Orthodoxy (Milbank 1990; Milbank *et al.* 1999) arose in opposition to what was discerned as a series of denials of key pillars of biblical truth by liberal Christians seeking to make their faith more palatable to the conditions of secular modernity. In contrast Radical Orthodoxy laid out an uncompromising vision of a socio-political role for the church in opposition to secularised capitalist society. The vision is founded on a series of understandings, of which four are key to this discussion (Milbank ND). First, the opposition of reason to revelation in secular modernism represents a corruption of Christian traditions that insist that human nature can only be understood fully with reference to supernatural destiny and divine disclosure. Second, without belief in God, humanity becomes nihilistic; nullity grows at the core of human being that regards death as more real than life, and abstraction becomes the truth and reality of the day. Third, although influenced by postmodern thinking, Radical Orthodoxy contests the idea that truth cannot be grounded in a certainty of intuitive presence, and thereby argues that contemporary efforts to value postmodernity are empty without a belief in God. Fourth, such belief needs to be outworked at both cosmic and communal scales, and Radical Orthodoxy encourages a 'theurgy' which is social, encouraging participation, and communitarianism on the basis of true Christian virtues.

According to Doak (2007), these theological foundations have been applied according to three rather different 'political ecclesiologies'. She suggests that Milbank himself envisages Christian religion as a kind of remnant Christendom in amongst a sea of secular reason; fighting back the waves of atomistic individualism and ontological conflict that so often seem to be resolved only by violent practices of governance that work against religiously-inspired roots of mutual participation and unity-in-difference. As a result, Milbank's political ecclesiology rests on an assumption that Christian theology should not enter into dialogue with contemporary (secular) social theory on the grounds that to do so would be to accept the assumptions therein about Godless social reason. Indeed, only explicit recognition of Christian ideas, for example, about forgiveness and reconciliation, can redeem society, and the key function of radical orthodoxy thus becomes to 'reclaim the world' by understanding it through 'a theological framework'

(Milbank *et al*. 1999, 1). This extreme and regressive form of postsecular commentary about how religion finds its voice again envisages a set of rather hazy boundaries between church and state, with the church positioned as the redemptive force for society and at best tolerating the state as a pragmatic necessity rather than as a partner in that process. Further, as 'the church' in this context clearly refers to Christianity, the approach appears to place other religions in the 'to be redeemed' category rather than recognising the possibility of inter-religious unity in postsecular contexts.

Other disciples of radical orthodoxy have developed this initial position. In what Doak (2007) labels a political ecclesiology of anarchic opposition, scholars such as Bell (2001) and Cavanaugh (1998) have offered a politically dialled-up version of what is necessary to bring the redemptive power of Christian values to bear in contemporary society. Their arguments are founded on a twofold critique of the contemporary state. First, the state has itself become an institutionalised personification of ontologies of violence and injustice, and cannot be relied upon to be the arbiter of common ground. It follows that the church needs to provide a radical alternative to this status quo of governance, provoking a deliberate witness to and testimony of the evils of the state. As discussed in later chapters of this book, there are clear signs in the UK and other contexts that the church is acting as a Habermasian 'community of interpretation', speaking truth to power over issues of inequality and injustice arising from overbearing policies of austerity in areas of welfare and care that disproportionately victimise the most vulnerable citizens in society. Second, and drawing on wider social theory from Foucault and Deleuze, the state governs using technologies that capture desires and align them with the disciplinary requirements of the capitalist machine. This binding together of governance with the capitalist economic system suggests not only a need for socio-economic transformation, but also a deeper requirement for the liberation of these desires – and here the role of spiritual therapy is advanced as a means of redirecting desire back to God through the unique polis and oikos of the church. For Bell and Cavanaugh, forms of Christian anarchy represent the only means by which these objectives can be advanced, and in turn this idea of religiously-inspired anarchic practice raises obvious questions about the scope for partnerships with fellow anarchists whose motivation is founded on other, non-religious, ideologies and normativities. Our discussion in Chapter 5 of the potential for 'movements' of postsecularity engages more fully with this question.

Perhaps the version of radical orthodoxy that comes closest to the idea of postsecular rapprochement is that expounded by Ward (2000) in his book *Cities of God*. In contrast to the separatist strategies discussed above, Ward advances a political ecclesiology based on critical engagement with secular modernity. Whilst retaining the familiar idea that Christian theology presents a truly radical critique of secular modernity and postmodernity, he insists that Christians should retain and develop existing relations of mutual participation in order to exert a positive influence on wider society. The broad task envisaged here is to bring about greater conformity between an idealised imagination of the city of God and the secular nihilism of the contemporary city, and there is a confidence that through critical engagement Christians can achieve a positive influence on wider society. In his later work, Ward (2005) fleshes out how a re-emergence of the sacred can involve direct influence on the institutions and cultural practices that shape public consciousness through a contextualised theology that deliberately pronounces on the 'signs of the times' (p. 3). This involves a form of religious engagement with society through involvement in public discourse in which a series of ethical values born of transcendent hope are culturally inscribed so as to contribute to the production of public truth. By this argument, there is an acceptance that religious truth cannot be held separate from the society in which it is performed; indeed, Christian values are already implicated – by their presence and by their absence – in the cultural negotiations of the contemporary city. Accordingly, Christian religion works from a position of already being in the world, and despite suggestions (Barnett 2017) that this positioning inevitably reproduces the expelling tendencies inherent in capitalism, this situated location permits Christianity to hone its capacity to demonstrate ways of life that are both critical of secularised ideologies, and present alternative embodiment of peaceful and redemptive values.

These different versions of post-liberal theology often fall short of providing clear illustration of how these discourses and practices might be enacted in the modern political world, and it is often assumed (despite the references to anarchist politics) that a 'natural' outworking of resurgent religion within contexts of postsecularity takes shape in ways described by key commentators such as Blond (2010) and Bretherton (2010a). Blond (1996) explicitly deploys discourses of postsecular philosophy in his diagnosis of how the infrastructure of civil society has been eroded; he bemoans the individualism and amoralism of secular capitalist modernism, and identifies a deep moral crisis stemming from an increasing inability to recognise what is objectively good, and what practices of virtue are necessary to achieve that good. His response is to argue for a reconstructed politics of virtue in which Christian values are revitalised as the arbiter of the common good. In practice, he fashions this vision into the persona of 'Red Tory', aligning a fundamentally right-wing critique of centralised state intervention and cultures of welfare dependency with advocacy for supposedly more leftist (but in reality, more culturally conservative) forms of communitarianism. In many ways, Blond's vision for the common good seeks to reinstate a more unitary culture by appealing to the supposedly cohesive societies of the past – a point that has not escaped his critics who find difficulty in translating such reinstatement into contemporary pluralist society. As Gray (2010) points out:

> In setting out his vision of the common good, Blond might have paused to consider the extent to which the more cohesive societies of the past were also more repressive. It is all too easy to succumb to Romantic visions of lost harmony. If you were a woman or gay, Jewish or a member of a minority Christian tradition, you might find the strong communities of medieval Christendom more than a little claustrophobic.

> (N.P.)

So, while there may be wider recognition of postsecularity in wider society, any perceived blurring of sacred/secular boundaries is likely to be understood in rather different terms than any lurching reversion to Christian orthodoxy. As Gray (2010) continues:

> Britain today is home to a plurality of religious traditions, ranging from varieties of theism through to the many strands of Hinduism and the godless spirituality of Buddhism. There are also many kinds of agnosticism and scepticism, some indistinguishable from undogmatic versions of faith. This rich and interesting diversity is one reason why Blond's project of reinstating a more unitary culture is so deeply problematic. Today there is no possibility of reaching society-wide agreement on ultimate questions. Happily, such agreement is not necessary, nor even desirable. No government can roll back modernity, and none should try. We may be in a mess. But the pluralist society that Britain has become is more hospitable to the good life than the imagined order of an earlier age, which in the end is just one more stifling utopia.
>
> (N.P.)

If Blond's backward-looking version of values-based politics is commonly viewed as 'tradition-centred, ecclesial, anti-modern theology' (Graham 2013, 120), Bretherton's account, which also shares a radically orthodox and postliberal heritage, benefits from practical involvement with the London Citizens in which representatives of different faith-based and secular interests worked together on issues of common local interest, such as the London-based campaign for a living wage. He uses this experience to advocate that Christian counter-performance of political and social relationships needs to be negotiated within a deliberate context of common life with others. Eschewing ideas that Christian faith needs to be kept as a separate 'remnant' of Christendom, Bretherton argues for partnerships of local community activism that involve a pragmatic hospitality between different religions and none. As evidenced by Jamoul and Wills (2008), the work of London Citizens was characterised by a commitment to shared goals that required some issues of religious, moral, and ethical difference to be set aside in order to focus on common objectives. For Christian participants, this necessitated a creative tension between remaining 'rooted in faith' while constructively 'breaking down the barriers between church and wider pluralist society'. In Bretherton's words, it involved 'stepping out of one's limited perspective' as a way of allowing 'new understandings to emerge' (2010, p. 87). While Bretherton is reluctant to embrace the discourse of 'postsecular', he argues that different constructions of religion, and therefore different constructions of secularity, are now bumping up against each other, and that these shared spaces of encounter provoke new forms of plural citizenship that reflect a mutual recognition that any particular position or worldview is not necessarily normal or normative for the public sphere.

These different outworkings of radical orthodoxy clearly demonstrate some of the complexities inherent in taking postliberal Christian theology out into the public sphere. In this specific example (and we move on to look at both other forms of theology, and other examples of how wider philosophy has implicated religious ideas for political purposes), very different mixes of discipleship and citizenship can be seen to emerge. These range from a separatist politics of redemption and anarchism, to more collaborative engagements in the public sphere that are inevitably shaped by the co-constituent partnerships involved. Here, the defining characteristics of postsecularity that we discussed in Chapter 1 enable some useful discernment in amongst these complexities. The extent to which processes and outcomes of religious involvement in the public sphere demonstrate receptive generosity, reflexively transforming rapprochement, and senses of re-enchantment will be a key factor in identifying political theologies capable of bearing the ethical fruit of postsecularity.

2.5 Alternatives to conservative political theologies of postsecularity?

The debates around radical orthodoxy illustrate how conservative appreciations of religious truth have fuelled multi-stranded struggles over how to turn that truth into political action. However, it is important also to consider how other – often far less conservative – forms of religious knowledge construction have conspired to effect a widely-recognised 'theological turn' in broader social and political theory. In a narrow sense, this reconsideration of religion has been a very gradual response to Schmitt's (1922) claim in the 1920s that all significant concepts in modern theories of the state are in fact secularised theological concepts. As Salter (2013) tells us, Schmitt, despite his unsavoury rightist politics, opened up a philosophical space for the possibility that theological ideas and beliefs will continually resurface within seemingly secular discourses (in his case regarding sovereignty and the force of law), regardless of the tendency for disenchanted politics of secularism to seek the exclusion of these charismatic ruptures in the natural and moral order. In a wider sense, however, the theological turn reflects a more far-reaching acknowledgement of complex issues relating to meaning, significance, and boundaries among politics, religion, and theology. Here we can recognise a deconstructed role for theology, that – in terms discussed by Lefort (1988) – recognises a nagging persistence within the structure of the political itself of an irreducible locus of transcendence. There is no suggestion here that religion and the political are indissolubly interconnected; rather that it is possible to recapture some of the inspiration that previous and contemporary thinkers have drawn from this imbrication of politics and theology.

These different approaches deal both in theological precepts, involving a process of unearthing dimensions of theological truth embedded within the body politic, and theological metaphors used to illustrate political potential (see Taubes 2004). Of major interest to the themes pursued in this book, however, is that in myriad contexts these theological politics affect within both secular and religious self-identifying parties a capacity for hope within what can otherwise be landscapes of ethical and political despair. For example, speaking from non-religious platforms, commentators of critical theory such as Eagleton (2010) and Critchley (2010) find hopeful possibility in the realms of religion and faith. In his discussion of Christian religion, Eagleton dismantles the pretensions of new atheists (see Dawkins 2007; Hitchins 2008) regarding their accounts of religious activity as flawed and ignorant. Admitting that the genuine radicalism of religion has typically been abandoned through processes of institutionalisation and accommodation to power, he nevertheless regards Christianity as having – in its original form – a potentially revolutionary role in contemporary politics. Here, he mirrors Connolly's (1999) earlier recognition of the possibility that religion can contribute new sources of ethical commitment to contemporary pluralist societies. Similarly, Critchley argues that religious ideas are still deeply embedded in politics, and that deep-seated religious notions such as faith, love, and hope retain an active capacity to breathe fresh air into current political situations. His expectation of faith is that it can provoke an urgent and active commitment to act in concert with others – a commitment that is insufficiently motivated by rationality alone. These hopeful claims are echoed across a range of theoretical fields, not least in feminism, where Braidotti (2008) sees the turn towards the postsecular as a challenge to European feminists to recognise that political agency and subjectivity can be conveyed through and supported by significant religious devotion and spirituality. This theme is put into practice in Olson *et al*'s (2012) study of how the postsecular is embodied in the lives of young people in Glasgow. In all of these examples, the concern is not to promote some religious revival, or even to achieve some more sophisticated grasp of theology, but rather to move towards new and more hybrid ways of understanding the political conditions we find ourselves in.

Underpinning this turn towards theologically hopeful narratives in politics are a range of quite different dialectic constructions, and for the sake of clarity we follow Cistelecan's (2014) threefold architecture to trace the varying forms and content involved. In Section 2.3 (above) we have already discussed his first category – that of liberal postsecularism. Habermas' (1995) debates with Rawls (1995) on reconciliation through the public use of reason, along with his subsequent constructions of the postsecular, are characterised here as finding a means of peaceable existence between secular democracy and religion; allowing religious people to join in the democratic debate by translating their religiousness into public discourses about values and practices. Cistelecan's verdict on liberal postsecularism is that it closes down rather than opens up possibilities for radical political alternatives:

> The basic problem with the formal formalism of the liberal postseculars is the same old structural problem of political liberalism: it is the fact that once a pure formalism is proclaimed as the political optimum, this form inevitably ends up by generating its own, exclusive content. Thus, what initially appeared as the most open-ended political arrangement, starts to operate as a genuine pensee unique, precluding any possible political alternative.
>
> (p. 12)

In other words (following Coles 2016), what seems to be on offer within liberal postsecularism is an identity to which people are expected to conform, rather than a more iteratively reflexive transformation into which all are welcomed. As we have already argued, we see this conclusion as being a little too pessimistic, not least given the propensity for hopeful spaces and subjectivities of postsecularity to bubble up despite, and in some ways because of, the adverse political condition. Thus, the plasticity of liberalism as the open space of plural political alternatives can sometimes permit dynamic political alternatives to emerge rather than simply submerging them in the saturating sea of endless possibility or forcing them into some new kind of hegemony.

Cistelcan's second category of theological politics can be understood as being influenced by post-metaphysical theology. The root of this category is planted in the soil of poststructural critique of both conservative and liberal theology. Rather than stridently reasserting (as do the postliberal theologians of radical orthodoxy) the traditional truths and values of religion as the normative guiding principles for dealing with the demise of triumphant secular reasoning, post-metaphysical theologians have recast the religious into a form in which the passing of the old God opens up opportunities for new religious projects. For example, Vattimo (2002 and 2009) sees social and political pluralism and the absence of metaphysical foundations on which to build process and structure as 'signs of the times' that characterise late modernity. With no single dominant lens through which to interpret the world, and with the triumph of interpretations over facts, he expounds Christian religion as an extension of broader philosophical ideas about 'weak thought'. Thus, for Vattimo, Jesus, with his ethic of kenosis (or self-emptying) becomes the instigator of a broader weakening that has culminated in the desacralising conditions of late modernity, such that faith in Jesus becomes a faith in the weakening of strong structures and dogmatic thought. In the face of nihilism, therefore, religion (after Levinas and Derrida) needs to seek safety in the Other, presenting ethics that are stronger than moral relativism, but weaker than normativity in their recognition of the situated and provisional nature of knowledge about the world.

Vattimo therefore advocates a 'weakening' of God, thereby allowing religious caritas to be understood as an embodiment of the weakening principle – a contesting of political violence through weakness in ways that cannot easily be secularised. His ideas are echoed by Caputo (2001) who expounds a postmodern interpretation of the weakness of God (1 Corinthians 1:29) refracted through a Derridian philosophy of hesitation that is open to weak forces relating both to the indeterminable nature of the other and to the unknowability of the future. In this way, post-metaphysical religion becomes – via 'weak theology' – a love of the unseeable: God is a weak force that notwithstanding still makes an unconditional claim on the lives of humans. Caputo therefore suggests that one significant contribution of faith is to point to a reality beyond the visible – a 'hyper-real' – which expands the narrow-minded horizon of possibility set by modernity. He argues that this hyper-real of faith is unlikely to be found in what he regards as the all-knowing, God-substituting certainties of fundamentalist religion, but rather in a more not-knowing kind of faith that relies on an endless translatability between the divine and the human, played out in love, beauty, truth, and justice. Thus, the love of God becomes integrally connected to the transformability of faithful lives, and to the transformative possibilities of the future. For Caputo, then, faith is literally an enactment that bears witness to the love of God – a leap of love into the hyper-real, a contradiction and reversal of culturally controlling drives, an unhinging of human powers and a drawing on invisible powers. Such a faith, acting as a gateway to invisible powers, and a pathway to seemingly impossible hope, connects well with critiques of secularism and new interest in spiritual landscapes (Dewsbury and Cloke 2009) in which a new sense of the sacred can be uncovered within the emergent postsecular. Caputo (2006) points to the 'anarchic effects produced by re-sacralising the settled secular order' and of producing an 'anarchic chaosmos of odd brilliant disturbances' which he sees as 'gifts that spring up like magic in the midst of scrambled economies' (p. 291).

When Caputo seeks to illustrate these leaps of love into the hyper-real he points precisely to the kinds of involvements of those motivated by faith in serving marginalised people that form the focus of our argument about emerging postsecular spaces in the city:

> If, on any given day, you go into the worst neighbourhoods of the inner cities of most large urban centers, the people you will find there serving the poor and needy, expending their lives and considerable talents attending to the least among us, will almost certainly be religious people – evangelicals and Pentecostalists, social workers with deeply held religious convictions, Christian, Jewish and Islamic, men and women, priests and nuns, black and white. They are the better angels of our nature. They are down in the trenches, out on the streets, serving the widow, the orphan, and the stranger, while the critics of religion are sleeping in on Sunday mornings.
>
> (Caputo 2001, 92)

While at least some of the work Caputo describes will be dutiful and morally selving, his significant contribution here is to identify the role of enchantment in these performances. By being willing to countenance the impossible, this faith-by-praxis spills out agape and caritas into situations of need and becomes attractive to others who from other motivations have reached the risky positionality of doing something on the basis that doing something is better than doing nothing. The weakness of God, in Caputo's analysis, becomes an expression of God's vulnerable love and faithful justice, serving as an inescapable contrast with the image of the violent sovereign warrior previously impregnated into theological politics. In Kearney's (2012) terms, this post-metaphysical theology points to anatheism – a faith beyond faith in a God after God; a new encounter with religious wonder, a more responsible way of encountering that which we think of as divine, and a discovery of the hidden holiness that constitutes everyday divinity.

Cistelecan labels his third category of contemporary political theology as Leninist messianism, a trend which he sees as 'a new founding of a revolutionary politics against both the pragmatic reasonableness of political liberalism and the hopeful resignation of post-metaphysical theology' (p. 14). In *Specters of Marx*, Derrida (1994) argues that any renewed leftist project must be reconciled with both the messianic promise implicit in Marx, and with what Judaic, Christian, and Islamic traditions have to say about messianic hope:

> The effectivity or actuality of the democratic promise, like that of the communist promise, will always keep within it, and it must do so, this absolutely undetermined messianic hope at its heart, this eschatological relation to the to-come of an event and of a singularity, of an alterity that cannot be anticipated.
>
> (p. 65)

In Christian traditions, redemption is only guaranteed by sacrifice, but in democratic politics, messianic thinking tends to defer the idea of sacrifice but retains the possibility of a 'second coming'. Key authors of messianic political theology, however, seek a theory of revolution; and their metaphorical use of Christian messianic apparatus accepts religion as an already-available conceptual dispositive of that revolution. Thus Agamben (2005) finds significance in the fundamental messianic texts of the West, while Badiou (2003) presents the biblical figure of St Paul as harbouring a genuinely revolutionary potential in contemporary politics. In this case, St Paul is regarded as a militant figure who serves as the poet-thinker who demonstrates fidelity to the Christ-event. Žižek (2001) sees St Paul as an engaged figure of struggle that cuts through liberal multiculturalism and pragmatic reformism, cultivating instead a community of believers glued together by a fidelity to a cause rather than by any other kind of master signifier. Widening out the scope of biblical reference, Hardt and Negri's (2001) exposition of empire likens potential revolutionaries of today to Christians of the later Roman Empire, witnessing the inexorable hollowing out of the terrestrial order of things, and the beginnings of a new, rejuvenating era of barbarian migrations.

Milbank (2005) translates these various biblical metaphors as evidence of how theology is being invoked in different ways by key thinkers in material socialism to visualise an appropriate ontology after secularism:

> Badiou sustains the possibility of a revolutionary event in terms of the one historical event of the arrival of the very logic of the event as such, which is none other than Pauline grace; Žižek sustains the possibility of revolutionary love beyond desire by reference to the historical emergence of the ultimate sublime object, which reconciles us to the void constituted only through a rift in the void. This sublime object is Christ.
>
> (p. 399)

However, Cistelecan insists that the focus in political theology according to Badiou and Žižek is not on the basic principles of a just society, such as grace and love, but on process of political strategy by revolution. If political theologies of postsecular liberation and post-metaphysical theology offer substance, the metaphorical use of Paul by Badiou and Žižek presents a Leninist (that is voluntarist and subjectivist) revolutionary form – the ritual of the messianic event. This suggests that some elements of postsecularity provide succour to those unbelievers who seem to require some kind of experience of belief involving elements (even metaphorical) of religious truth and ritualistic frameworks in which they can believe; what Critchley (2012) refers to as the politics of mystical anarchism.

There is a significant risk that these three frameworks of political theology at the heart of the theological turn in political theory all point to a decentring of the role of religion in the public sphere. Liberal postsecularism risks being denuded of its core religious content through both the requirement to translate itself into secularly dictated discourse in the public sphere, and by continuing pressures to practice foundational values in private. Post-metaphysical theology trades on the deconstruction of metanarrative, including that of God. Its enchanting appeal relates to faith in the unseen, and while this can be regarded as crossing over into more traditional terrains of religion involving the prophetic and the charismatic, its inherently weak theology disavows any sense of normativity beyond the provisional and the situated. What Cistelecan dubs as Leninist messianism represents an attempt to rescue what is supposed as the progressive elements of rational modernist enlightenment, using biblical metaphor to satisfy the need for sustained religious ritual in amongst mystical anarchism. Here, too, the theo-ethics of religion are sidelined in favour of their metaphorical form. Big theory, then, seems to be playing a complex and multifaceted hand in its theological turn, and in so doing may provide less direct encouragement for conditions of postsecularity than might otherwise be assumed.

2.6 The messy middle of postsecularity

Thus far in this chapter we have presented bodies of conceptualisation that provide different philosophical containers for debates about the postsecular. Some of these debates seek to draw on grand theory to support arguments about the vacuity of postsecular ideas, and it is easy to recognise the difficulties in identifying clear postsecular territory if the discussion is founded on binaric principles of secular versus postsecular. As this chapter shows, just as supposed binaries between modern and postmodern are of little use because 'we have never been modern' (Latour 1993), so any imagined binary between secular and postsecular needs to be dismantled because 'we have never been secular' (see Ward 2014). To stage a debate about postsecularity at this binaric scale seems doomed to failure, and recent attempts to revise the secularisation thesis point us towards the interstices of secular/postsecular rather than to their separate heartlands. We could at this point, as some have done before us, simply give up on the idea of the postsecular – were it not for an empirical conviction that something is 'going on out there' that both requires new kinds of explanation and deserves attention for its potential for new forms of ethical politics. Therefore, reports of a change in public consciousness towards religion, and potential technologies for ethical and political rapprochement in crossover and assimilative narrative and action seem to us to be offspring worth saving from the emptying out of the grand theoretical bathwater. Postsecularity offers an appropriate heuristic approach involving the kinds of technologies, tools, and vocabularies that are needed to make sense of how the religious and non-religious co-exist in ethical and political life.

Even so, postsecularity remains stubbornly hard to pin down. When postliberal theology occupies the philosophical space, adherence to normative religious truths leads to very different political ecclesiologies; suggesting the possibility both of an isolated religious remnant seeking to redeem the world either anarchistically or through the straightforward power of example, and of groups of religious people engaged in a citizenship of mutual participation. Although such participation can also take anarchistic forms, it is most often characterised as a rather conservative and non-radical response to surrounding political circumstances. Meanwhile, from post- and non-metaphysical stances, the turn towards religion seems to be dependent on a weakening or even dissolution of religious normativity in order to pursue either a messianic metaphor for political revolution, or a kind of weak mysticism that makes occasional donations to the goodness of people's lives. Nevertheless, we retain more than a sneaking suspicion that in amongst these redemptive ambitions, participatory practices, revolutionary intentions, and mystic disturbances there lies a potential for receptive generosity and rapprochement between self and other that can contribute significantly to the re-enchantment of contemporary societies (see Chapter 3 for further discussion of this). In order to pursue this possibility, we advocate the use of other, more modest, philosophical territories on which to conduct the postsecular debate.

In their essay on different modes of attentiveness relating to homelessness, May and Cloke (2014) seek to identify important strands of understanding that have been obscured or devalued by previous conceptual orthodoxies. Invoking Gibson-Graham's (2006) 'reading for difference', they argue that previously dominant ideas about neoliberalism and revanchism have resulted in only partial understandings of the landscape of homelessness, and that alternative narratives – embracing ethical motivations and actions that run counter to this punitive framework – provide an alternative emphasis on emancipatory participation in responses to the needs of homeless people. Their approach embraced two sets of logics:

> First the alternative interpretative potential within overarching grammars of revanchism, notably in terms of understanding some homeless service spaces as embodying resistance rather than co-option by a neoliberalising welfare system; second, the emergent postsecular discourse and praxis that signal potentially significant changes in the secular-religious relationships manifest in a greater willingness for cross-boundary partnership in order to

'do something about' homelessness in ways that oppose the grammars of revanchism.

(pp. 894–895)

In so doing they identify a 'messy middle ground' (see Murphy 2009) between the pessimistic narratives of homelessness based on the punitive conceptual grammars of neoliberal revanchism, and the admittedly more optimistic potential of activities whose postsecular characteristics become visible in the more humdrum spaces of care and welfare in the city. Both sets of grammars are important, but the key message is that reliance on the precepts of big theory can cloak the potential for innovative ethical and political participation that takes place in the mundane and ordinary spaces and subjectivities that dwell at ground level. Their claim is that despite presenting some secular academic thought with an unwanted conundrum (that is the possibility of a progressive left-leaning rapprochement between religion and other political activity) attention to these ordinary spaces of care yields significant examples of resistance and resilience in the homelessness sector. But this innovation is only likely to be recognised by those who are capable of being attentive to it.

This plea for attentiveness to what are often small-scale and subtle circumstances and events represents an important challenge to those whose instinct is to reject the idea of postsecularity on macro-conceptual grounds. Recent research into the phenomenon of food banks has noted a similar discontinuity between the macro-political response that food banks should not exist (given the responsibility of the state to prevent food insecurity) and local-scale responses to need 'in the meantime' (see Cloke *et al.* 2017; Williams *et al.* 2016) which often involve ground-level spaces of participatory potential which remain invisible to the macro-perspective. So, while it remains crucial to pursue big concepts of secularity and political theology, we also need to recognise the possibility that key spaces and subjectivities of postsecularity may often resist and defy being captured by these concepts. In the following chapters, we proceed to examine these subjectivities and spaces in more detail.

3 Subjectivities

3.1 Emergent subjectivities of postsecularity

This chapter examines the formation of ethical subjectivities that emerge in and through postsecularity. While critics of postsecularity suggest its novelty is more as an intellectual framework than as an empirical phenomenon, we want to foreground in this chapter the seismic empirical shifts in subjectivity that have led to the rupturing of previously calcified lines drawn between the 'religious' and 'secular'. In particular, we emphasise the emergent subjectivities that seem to be bubbling up through this process, and the ways in which they are reterritorialising the contemporary landscape of religion, ethics, and politics. Building on arguments made by Cloke and Beaumont (2013), we suggest both the initial willingness to enter into the spaces of postsecularity, and the subsequent prevalence of such entries, have been intensified by the erosion of religious and secular fundamentalisms. We seek to unpack this process further in this chapter by theorising the emergence of postsecularity and its subjectivities in relation to the existential-phenomenological life produced in neoliberal late capitalist societies. We pursue this argument in three sections. Section 1 outlines the changing constitution of subjectivity under late-capitalism and how disenchantment with neoliberal 'myths' (Barthes 1972 [1957]) has created a void in which has grown a complex amalgam of individualism, charitable sentiment and ethico-political weariness. In Section 2, we outline the ways in which subjectivities of postsecularity can be considered a by-product of, and response to, this political-economic and existential set of conditions in late capitalism. Attention is given to the ethical and political possibilities presented by postsecularity, especially in the releasing of ethical energies and values that assist in the reformulation of religious and secular subjectivities. Section 3 traces these emergent subjectivities of postsecularity across three case-studies. Our purpose here is to illustrate the ways in which postsecularity in different contexts serves as a catalyst in generating energies for receptive generosity, rapprochement and hopeful re-enchantment. We begin, however, with a brief review of geographical understandings of subjectivity, which is followed by a discussion of what is meant by the idea of 'subjectivities of postsecularity'.

Subjectivity in general terms refers to a person's thoughts, feelings, desires, attitudes, hopes, and consciousness. Across the social sciences, the notion of subjectivity has been addressed using a multitude of theoretical perspectives. Early articulations of Humanist geography, for instance, emphasised a self-conscious and reflexive subject, thereby positioning the human subject as the centre of identity (Doel 1995). The cultural turn in the 1990s, and in particular engagement with the work of Judith Butler (1990), Michel Foucault (1980) and others, popularised social constructionist approaches to subjectivity characterised by emphases on performativity, discourse, and interpellation. Subjectivity in these terms was thought to be produced through language, conceiving identity as the inscription and repetition of discursive practices. Feminist, non-representational, psychoanalytical, and actor-network theory approaches, albeit in very different ways, have turned from using identity as a category. Instead, there has been movement towards theorising the embodied and affective aspects of subjectivity that challenge both Humanist notions of the intentional conscious subject and the primacy given to discourse associated with early social constructionist approaches.

Non-representational and post-phenomenological approaches conceive subjectivity as something unfinished, provisional, emergent, and potentially open to change (see Simpson 2017 for an excellent review). Non-representational theory (NRT) is characterised by a 'concern for the practical, embodied "composition" of subjectivities' (Anderson and Harrison 2010, 8). In earlier articulations of NRT, empirical foci demonstrated the material-situatedness of bodies within space (Thrift 1996; McCormack 2014), while recognising that embodied and affective encounters are relationally configured by, and have the excessive capacity to disrupt, social formations (Dewsbury 2007). More recently, theorisations of affect, desire, and the event have provided new insights into the workings of neoliberal politics and capitalist political-economy (Anderson 2012 and 2014), highlighting how affective life is organised and mediated by forms of power and embodied dispositions.

In this book we adopt a broadly poststructural perspective that is attuned to the more-than-representational aspects of space, politics, and subjectivity. Such an approach helps foreground the co-constitution of religious and secular subjectivities and offers a categorisation not of unitary bounded subjects, nor of mere discursive entities, but of fully alive and embodied beings whose boundaries are porous, unfinished, and becoming. Taking this approach to an analysis of postsecularity is helpful on two grounds. First, as an analytic device, it focusses on the emergences and fragilities of subjectivity – on its potentialities rather than its fixity – thereby giving analytical space to often overlooked resonances and dissonances that push against and evade easy categorisation as 'religious' and 'secular'. Such an analysis pays greater attention to the spaces in which subjectivities are produced, reworked, and challenged. Subjectivities of postsecularity, we argue, are not pre-existing or stable categories, but are in the process of becoming, enfolding, and constantly in flux. The ethical capacities of such subjectivities are not guaranteed but are produced through *relationalities*, which include heterogeneous material and immaterial elements, encompassing discourses, practices and performativities, affective atmospheres, and ethical deliberation (Holloway 2013). This foregrounds the constitutive power of space itself in configuring the possibilities – and barriers – for subjectivities of postsecularity to emerge (see Chapter 4).

Second, such a perspective reveals the liminal movements in-between extant categories of the religious and the secular. Our argument is that postsecularity is a thirdspace that people enter into, dwell in (albeit often temporarily), and leave (Baker 2009); a space in which ethical sensibilities of receptive generosity are formed and reaffirmed, and which can ultimately be productive of new ethical and political sensibilities. Postsecularity is attuned to the potential for partnerships across religious and secular actors to reterritorialise expressions of religion and secular belief, identity, and practice. The ethical values generated in postsecularity go beyond a simple expression of theo-ethics as visible in the work of FBOs. Rather they are characterised by hybridity, produced out of an explicit 'crossing over' of religious and secular narratives, practices and performances that are revealed in particular spaces of care, welfare, justice, and protest, and in particular subjectivities that reflect a commitment to in-commonness, and a progressive responsibility to reach out to alternative notions of common good.

For definitional clarity, then, we argue that *subjectivities of postsecularity* refer to the affective desire for a greater degree of in-commonness and the cultivation of sensibilities that prioritise 'common good'. These sensibilities are reflected in, and produced by, *ethical capacities for receptive generosity, rapprochement and re-enchantment.* These characteristics are not always intentionally scripted or designed, but are developed in situ, in relation with others, and shaped by a confluence of different ethical, theological, and political flows and motivations. For this reason, we refrain from using the term 'postsecular subjectivity' which implies a relatively stable subject position and invites an interpretation of postsecularity as a deliberate and organised process. Again, we should emphasise here that notions of the common good can represent a guise for conservative politics, an ignorance for identity politics, and provide moral rationalisation for exclusion. Yet the stance towards in-commonness advocated here is not a utilitarian ethic that tolerates the logic of sacrifice of some people/ places to appease some 'greater good'. Our conception of being-in-common seeks to avoid any normative blueprints and does not serve as an idealised description of the current state of affairs. Rather, it refers to ethical stances that allow more generous identification to emerge across difference. Inspired by insights from human geography (Popke 2009) and public theology (Volf 1996), we ask what kinds of selves – and self-other sensibilities – do we need to be in order to recognise interdependencies. Being in-common therefore denotes emotional and physical connectedness to the 'other', a process of moving outside of the normal habitus. Hence receptive generosity refers to the making of space within the self for the other and must be dialogical and sustained through engagement with the other. Otherwise it can become a consolatory posture that assures the self of its radical receptivity while allowing creeping distance to

harden in-common sensibilities and practices. Equally overfamiliarity or proximity can produce scripted responses premised on the logic of assimilation of the
other into the 'thought prison' of the Same (Doel 1994) or generate blasé attitudes to difference that circumscribes meaningful embrace (Volf 1996). We
argue that ethical capacities for receptive generosity, rapprochement, and hopeful
enchantment – the defining characteristics of subjectivities of postsecularity –
are both circumscribed by, and summon a response to, the decaying excess of
neoliberal subjectivity.

3.2 Neoliberal subjectivity: late capitalist desire, ressentiment, and enchantment

Accounts of postsecular rapprochement have been developed largely in relation
to how neoliberalism has created opportunities for faith groups to expand their
public involvement in service provision, care, and advocacy (Beaumont and
Cloke 2012). While this approach highlights the deeply contested terrain of neoliberal governance in creating spaces of postsecularity, we want to elaborate on
this argument by contextualising the emergence of postsecularity as occurring in
the deeper fissures of modern existential and phenomenological life that shapes
the logic of acquiescence and resistance. Accordingly, we contend that an understanding of the emotional and affective conditions of late capitalism, and the formation of desiring subjects through neoliberalism is key to an analysis of
postsecularity. In what follows, we deploy a series of conceptual frames that for
us seem to encapsulate key processes working on religious and secular subjectivities, shaping the contemporary landscape of ethics and belief.

Regimes of desire and precorporation

While much scholarly attention has puzzled over the strange 'non-death' of neoliberalism (Crouch 2011, viii), pointing to the durable lobbying power of multi-
national corporations in deforming democratic politics, it is equally important to
assess the ways in which late capitalism and neoliberal logics are produced, cir-
culated, and sustained through an amalgam of emotion, belief, and affect (Vrasti
2009; Anderson 2016). Following Lordon (2014), we argue that capitalism must
'be grasped not only in its structures but also as a certain *regime of desire*' (cited
in Rose 2017, 248). We unpack this argument as follows. In recent years neoliberalism has been commonly understood through an analytic of governmentality, a framework which has helped focus analysis of the mundane practices
through which neoliberal spaces and subjectivities are produced at a range of
different scales (Larner 2000). This perspective has opened up renewed attention
to the performative and affective workings of neoliberal capitalism which has
penetrated emotional and affective life. We are not exclusively referring here to
the discursive construction of the rational, calculating, and entrepreneurial
subject who self-regulates their behaviour towards self-betterment, or wider
neoliberal dispositions that valorise the so-called virtues of acquisition and

competitiveness that characterise the possessive individual. We also seek to acknowledge the importance of emotional and affective registers as sites of capitalist inscription – working through the freedoms, emotions, affective labour (having the 'right attitude') – and establishing the possibilities through which we narrate our relation to life itself. As Derrick Jensen (2006, 552 cited in Tim Jenson 2011) explains:

'It would be a mistake to think this culture clearcuts only forests. It clearcuts our psyche as well. It would be a mistake to think it dams only rivers. We ourselves are dammed (and damned) by it as well. It would be a mistake to think it creates dead zones only in the ocean. It creates dead zones in our hearts and minds. It would be a mistake to think it fragments only our habitat. We, too, are fragmented, split off, shredded, rent, torn'. When these territories of desire and imagination are stolen, ravaged, and toxified it becomes that much easier for the theft and destruction of natural landscapes to go uncontested, unnoticed.

Tim Jenson underlines that the hegemony of neoliberalism is not simply maintained through a common economic logic, but through a 'common sensorium'. He argues:

[n]eoliberalism entrains us to experience certain emotions over others, suggests rules for their expression, and even tries to define what one is 'allowed' to feel for. These everyday flows of feeling – from bodily intensities of relation (affect) to their narrativised accounts (emotion) – habituate us to the cadence of neoliberal subjectivity.

(Jenson 2011)

It is in this sense that 'the hegemony of late capitalism is not fought on rationality; but rather on the territories of the personal, the affective, and the aesthetic' (Vrasti 2009, 3). Affective competencies of late capitalism are accompanied with *cruel optimism* – the endurance of pain and exhaustion in the pursuit of the better life, despite the desired future remaining unattainable and its pursuit often entailing self-destruction (Berlant 2011). Underpinning this is the fusion of *immaterial labour exploitation* (namely, the extraction of surplus value through both emotional labour and the compulsion of unpaid and 'take-home' work), and *material precarity* (the realisation that the 'good life' as idealised in homeownership, pension security, and secure employment is no longer attainable for the majority of the younger population in the UK). Paradoxically it seems, emotional life in late capitalism is becoming increasingly *automated* by technologies of self-regulation, targets, and a normalised gig economy in which 'get the job done' mentalities have come to legitimise sequential servitude to the 'next task' and downgrading of working conditions.

Highlighting prominent elements of the existential conditions in late capitalism helps reveal the ways in which desire is shaped by capitalist rhythms and regimes of immaterial labour. In recent work (Cloke *et al.* 2017), we have discussed this in reference to the idea of 'precorporation' which refers to 'preemptive formatting and reshaping of desires, aspirations and hopes of capitalist culture' (Fisher 2009, 9). Fisher uses the work of Deleuze and Guattari to understand capitalism as axiomatic in the ways which it 'seamlessly occupies the horizons of the thinkable', seeping into and colonising the 'dreaming life of the population' (2009, 8). As a result, capitalism hollows out or absolves religious and nonreligious belief. Belief itself, Fisher argues, has been 'collapsed at the level of ritual or symbolic elaboration, all that is left is consumer-spectator' (2009, 4). Desire, in Deleuzian terms, is not a negative or excessive entity to be controlled, but rather an 'affirmative vital force' that is productive and actualisable only through practice (Gao 2013). We want to argue that the kind of production of desire in late capitalism warps our relation of self-to-self and self-to-other, generating a deeply ambivalent ethical landscape.

Spiritual ennui and existential ressentiment

Such an ethical landscape, carved from the affective burdens of neoliberalism, has sapped emotional resources for hopeful politics. Scholars, from a variety of philosophical and political commitments, have converged on this discursive space, trying to discern what can be described as the 'spiritual ennui'[1] of neoliberalism – a weariness towards social change and an emptiness of hope. From the onset we want to acknowledge the connection between modernity and ennui – in its guises of boredom, sadness, or melancholy – is not an original observation, and is found in various writings of Kierkegaard, Baudelaire, and others (see Ferguson 1995; Kuhn 1976; Radden 2000). As discussed in Chapter 2, we note the clear politics inherent to 'writing' existential culture, and we are wary of deploying ennui as a problem that requires a Christian ontology as suggested by proponents of Radical Orthodoxy (see Milbank 1995; Blond 1995; also see Martinson 2013). Instead, we use the term as a lens to think through the politics of affective life in late capitalism which, we argue, has had significant implications on the existential cultures of belief, including non-religious and religious expressions. Ann Cvetkovich (2012) and John Schumaker (2016), for example, note the escalation of depression and hopelessness that characterises the affective culture of individualism, materialism, debt, and overwork; while others have explored the political economy of unhappiness (Davies 2011), suggesting demoralisation, loneliness, and ill-health are amplified by neoliberal and late-capitalist regimes which produce toxic cultures defined by alienation from others, and ultimately the self. Schumaker (2016) and Monbiot (2017) claim that levels of intimacy, communal trust, and deep friendship are being exhausted, resulting in a 'psycho-spiritual crisis' in which victims feel 'disoriented and unable to locate meaning, purpose and sources of need fulfilment' (Schumaker 2016 para 5). Equally, Human Geography has been animated by recent debates about affective life, desire, and ethics, and the ways in which these connect to political-economies of neoliberal capitalism (Anderson 2016; Olson 2016). The grip, tenacity and seduction of neoliberal reason has been seen to produce both 'compliant' subjects, allegedly able to deal with the anxieties of precarity (Firth 2016; Anderson 2016), but also particular 'subjective forms (i.e. modes of conduct, affects, attitudes, social relations, and lifestyles) that are congruent with the capitalist logic of accumulation and competition' (Vrasti 2009, 7).

Theorising neoliberalism as spiritual ennui helps foreground the limiting factors that act as a potential bulwark against the emergence of postsecularity. The growing tide of meanness towards forms of 'dependency' is usually articulated with recourse to media stereotypes of 'scroungers', 'opportunists', and 'free-loaders', punitive government policy, and changing public attitudes surveys that show a hardening towards poverty. However, what is missing in this analysis is a recognition that these trends are not exclusively an outworking of a discursive interpellation. Behind the discursive actualisation of these attitudes resides a more troubling development – *existential ressentiment* (Connolly, 2008), understood as the assignment of blame for one's frustration onto others – which can encompass attacks on those who *appear* to 'have it all' as well as meanness towards calls for justice or ameliorative agendas because of a sense of being wronged oneself. Recognition of self-sacrifice is thus transferred onto an expectation for others to do likewise. It is these forces that legitimise a so-called compassionate indifference where care is apportioned according to a metric of trustworthiness and deservingness. At the heart of this, we argue, is a closing off of the self and ignorance towards the demand of the other; a non-recognition of distancing oneself that produces an affective collective experience that resonates with personal and collective guilt. This guilt can prompt a longing for restitution – either in the form of pro-justice gestures, or more radically receptive stances of being-in-common. Desire for restitution can also be manifest in the avoidance of uncomfortable encounters, or worse, the will to expel undesirable encounters with others through displacement and containment.

We therefore argue that affective and spiritual life in late capitalism is a key site in the politics of postsecularity. Theorising neoliberalism as spiritual ennui reanimates the deeply contested terrain of the personal, affective, and aesthetic (cf. Vrasti 2009), drawing attention to the visceral dispositions and the formation of prepolitical desires that underpin political mobilisations (Connolly 1999). Crucially, it also foregrounds the political potential of ethical desire that deviates from, or subverts, neoliberal technologies of desire, suggesting that biopolitical life, affective competencies and immanent relations represent key sites of reproduction *and* resistance. Scholars have noted how this (de)formulation of desiring subjects through technological saturation, immaterial labour, and mythic narratives of aspiration, fairness, and happiness can be understood through the lens of disenchantment and enchantment, and it is to this we now turn.

Neoliberal enchantments

Neoliberalism also works through the dynamic of enchantment and disenchantment, simultaneous processes swirling to shape late-capitalist subjectivities. The notion of enchantment, despite its disparate meanings, is useful for sharpening analysis of late capitalist subjectivities on two grounds. First, it helps us focus attention to ambiguous or bogus enchantments produced through neoliberal 'myths' (Barthes 1972 [1957]) of success, happiness, and meritocracy, which refract into embodiments of cruel optimism (Berlant 2011). In a context of the pseudo-sacralisation of acquisitionist values, and the mysticism of market self-regulation, the notion of enchantment also speaks to the speculative ontologies of financialism that operate on chance, contingency, and luck (Hilgers 2011). This recognition challenges dominant sociological narratives using Weberian notions of disenchantment to characterise the transition from the enchanted universe of medieval Christendom to the rationalised and taxonomised cosmos of modern Europe. Indeed, recent scholarship uses notions of enchantment to highlight the ways in which regulation of desire is not a disenchanted affair *per se*, but rather is saturated with, and works through, mystery, sacralisation, and enchantment (Rose 2017). One of the ways enchantment is manifest in mass culture is through commodity fetishes, which Braidotti (2008, 12) calls 'sacred monsters' of global consumption. She observes in late capitalist globalised economies, the 'digital saturation of our social sphere by fast-circulating visualisation technologies, the mystical overtones of global icons and the semi-religious cult following' (Braidotti 2008, 12) whereby modes of consumerism have become sacralised and material objects invested with magical powers (McCarraher 2005 cited in Cavanaugh 2014). For Cavanaugh (2014), therefore, secular modernity remains enchanted as the sacred is refracted onto political state-power and consumerism.

Second, understandings of enchantment and disenchantment helpfully hone our focus on the contemporary landscape of democratic politics. Political-economic theorists of the post-political (see, for example, Wilson and Swyngedouw 2014) warn of increasing political apathy for mainstream party-politics with its ritualised choreography of supposedly democratic electoral procedure. As neoliberalised capitalism proceeds with the transformation of nature and the appropriation of its wealth, government appears to become reduced to an associated bio-political management of desire and happiness. As a result, there is a risk of a broadly disenchanted paralysis of empty nihilism in which the only discernible telos lies in the individual pursuit of pleasure. Not only does this reduce the likelihood of encounters with less privileged others, but it actively feeds the neurosis of autonomous desire (Reinhard 2005) that seems to fuel stigmatising stereotypes of others-as-enemies. Moreover, as Connolly (2008) emphasises, some forms of religion have been directly implicated in this disenchanted regime of neoliberalised and austere capitalism; in his terms the Evangelical-Capitalist Resonance Machine has not only shored up the politics of existential resentment that fuels inequality and socio-cultural enmity, but has also contributed to active forms of disenchantment via the dogmatic espousal of desire based upon individualised prosperity and extreme moral conservatism.

These factors shape the values and expressions of both religious and secular actors. The desire for (re)enchantment is considered one response to the late-capitalist condition of digital saturation, immaterial labour, and the alienation economy. Neoliberal technologies of desire provide the narration for, and channelling of, the search for 'authenticity', autonomy and freedom of expression and individuality. Religious subjectivities are clearly not immune from these shifts in regimes of neoliberal desire and it is important to highlight the resonances and dissonances between late capitalist enchantments and religious enchantment. Possamai (2017), for instance, suggests that certain religious enchantments can embody features of 'hypo-consumerism' (p. 5) and warns against a hollowing out of collectivist religious precepts towards an individualised spirituality characterised by the urge on developing, healing, and entertaining the self, and the McDonaldisation of religion through commodified branding and standardisation. Individualist narcissisms that characterise late-modernity supposedly have also seeped into religion shaping its performativities and experiential self-reference, which some suggest is caught up in anxious confirmability surrounding belief and the production of Christian culture, while inversely works to temper radical receptivity to difference (Rieger 2001 cited in Barrett 2017). However, it is important here not to overlook the cracks and fissures in the political-economic moment, and the divergent possibilities for resistance as well as acquiescence in a range of familiar and expected places.

By emphasising the formation of neoliberal subjectivity and its affective registers, our argument is not to overaccentuate the successful capture of self-interested and entrepreneurial subjects; neither are we implying that self-interest and entrepreneurial spirit are mutually exclusive from charitable endeavour (see Muehlebach 2012). Rather, we highlight a series of fissures in modern existential and phenomenological life that have shaped the logics of acquiescence and resistance. It is in these cracks that energies are released which prompt people to detach from right-leaning religion and the enchantments of late capitalism. Resurgent reactionary sensibilities also grow in these cracks too, evidenced in the recent white supremacist marches in Virginia, USA, and the growth of neo-fascist groups in Europe. Emergent postsecularity is one mode of collective embracing of the horror of 'the powers' and neoliberal disenchantment together; yet as Chapter 5 demonstrates conservative or reactionary responses to this horror converge on the ego-move (and financially prudent strategy) of fascism. Neoliberal disenchantment therefore opens out a deeply contested terrain of ethical and political subjectivity in which emergent postsecularity should be considered a nascent opportunity rather than a certainty in these cracks.

3.3 Possibilities of postsecularity

How, then, has this political economic and affective context of neoliberal capitalism shaped the emergence of postsecularity? We want to argue that postsecularity is a by-product of, and response to, both existential life in late capitalism and the ways neoliberal enchantments have shaped religion and secular subjectivities. Following Cloke (2011a) and others (Cloke and Beaumont 2013; Williams 2014), we suggest that one key development in the emergence of subjectivities of postsecularity has been a desire for a counter-ethics in response to growing recognition of the excesses of neoliberal political-economy: the deepening levels of poverty and intensified precarity; the growth of corporate power; the conscious cruelty of 'welfare reform' and the pernicious injustices of racism, homelessness, and food insecurity that are so often met with wilful ignorance from government ministers (see Slater 2014). Neoliberal excesses play out in the material, social, and emotional burdens stacked on the backs of the poorest, as well as in the repugnant celebration of material wealth among the super-rich (Sayer 2015). We argue that it is the phenomenology of need understood as an intuitive, visceral, and embodied response to 'do something' in the face of injustice that underpins new energies for rapprochement and prompts religious and non-religious citizens to put aside moral or ideological differences to work together towards socially transformative goals. Such impulses are galvanised by, and find expression in, the desire to step out of idealised rhythms and enchantments of the neoliberal subject-citizen. In what follows, we discuss four interconnected possibilities offered by postsecularity as it emerges from this political economic and existential context.

Postsecularity as opportunity in neoliberal austerity: movements beyond religious and secular fundamentalism

In responding to neoliberal austerity and the associated phenomenology of need, subjectivities of postsecularity have been associated with moves away from prevailing religious and secular fundamentalisms, and towards experimental and hybrid formulations of religious and secular identity, belief, and practice. In what is now a familiar story, ongoing privatisation and retrenchment of the welfare state has created opportunities for faith groups to expand service provision roles, either directly by working in partnership with state apparatus, or indirectly, by filling the gap that the state has chosen to excise completely from the palette of social welfare (Beaumont 2008b; Williams *et al.* 2012). Within this narrative, important distinctions need to be made between different periods of neoliberal governance in the UK, with the New Labour government's relationship to the Third Sector representing an explicit recognition of the strengths of faith-motivation, incorporating religious capital for resource, cohesion, representation and provision (Baker and Skinner 2006). Since 2010, the rationale for partnership has changed considerably to reflect a morally conservative form of communitarianism that characterised the 'Big Society' and greater expectations have been placed on Third Sector groups to take on responsibility for roles and services previously provided by the public sector (Williams *et al.* 2014). In what has been coined 'austerity localism' (Featherstone *et al.* 2012), this period has witnessed more extreme forms of 'hollowing out' of public service delivery through ideologically driven disinvestment in local government, welfare, education, health, and social care funding – resulting in a geographically uneven and highly variegated welfare landscape characterised by privatisation and outsourcing to an array of different non-profit, for-profit, and voluntary organisations. English councils lost 27 per cent of their spending power in the period 2010–2015 with a further 56 per cent grant reduction in the period

2015–2020 (HM Treasury, 2015, 3 in Hastings *et al.* 2017). Unsurprisingly, therefore, the capacity of local government to meet the increasing needs of its citizens has been squeezed, resulting in (sometimes reluctant) partnership with or reliance on faith-based and other Third Sector involvement in areas previously provided for by the state. In this way, postsecular rapprochement has often been dismissively interpreted as religious actors being forced to water-down distinctive elements of faith in order to gain credibility and acceptance in the public realm; a subject-position constructed by government pressure on religious groups to self-regulate their behaviour as part of a domestication and privatisation of public religion (Bretherton 2010b).

However, in these circumstances, and sometimes despite them, faith groups have created new opportunities for faith-secular partnerships in areas of homelessness, refugee support, care and advocacy, and the like (Beaumont 2008b; Cloke and Beaumont 2013; Williams *et al.* 2012). Here, faith groups are both translating faith ethos into the public sphere and opening up their staffing and support base beyond faith communities. Rather than being a straightforward neoliberal by-product, then, faith-based activity can be recognised as a key protagonist in the emergence of postsecularity – by providing practical devices for rapprochement and spaces of reflexive translation of religious and secular positionalities, which in some cases, has brought about a change in the national and local states' 'secularist self-understanding' (De Vries 2006, 3).

Not only do these interstitial spaces of care and ethical response actualise incommon sensibilities, they also represent an opportunity for wider shifts in secular and religious consciousness to be cultivated and expressed in ways that encourage a willingness on the part of participant actors to eschew secular and religious fundamentalisms in order to enter into rapprochement with others. While Chapter 2 documented the movement beyond secular fundamentalism on an intellectual and philosophical register, we want to develop discussion of two additional currents of practice that have shaped secular capacity to enter into rapprochement. The first refers to the divergent, fragmented, and fluid constellation of belief itself; what Charles Taylor calls the 'supernova effect' – the splintering of positionalities accompanying the re-organisation of transcendental belief onto an immanent plane. Taylor (2007) notes that the new conditions under which choices for belief and unbelief are inherently fragile and hybrid in formation:

> We live in a condition where we cannot help but be aware that there are a number of different construals, views which intelligent, reasonably undeluded people, of good will, can and do disagree on. We cannot help looking over our shoulder from time to time, looking sideways, living our faith also in a condition of doubt and uncertainty.
>
> (Taylor 2007, 11)

Such doubt has diversified and blurred previously apparent certainties in belief and personal practice. In the turn to enchantment, for example, the mystical and the spiritual help to embody a postsecular space characterised by the deconstruction of fundamentalist binaries – religious/secular, rational/superstitious, sacred/profane, and so on. We argue that this blurring of binaries has contributed to practices of receptivity among disparate subjectivities, although there is no easy correlation between these spiritual sensibilities among the non-religious and their willingness to engage in the reciprocal generosity in postsecular partnership. Indeed, it might be the case that some forms of secular enchantment deliberately eschew postsecular rapprochement due to perceived metaphysical differences or identity politics that circumvent collaboration. Equally, it might be the case, as Carroll and Norman (2016) note, that commonalities are recognised across different constellations of (non)religious and spiritual expression, to the extent that someone engaged in Occult and Spiritualism might have more in common with Pentecostalism and Charismatic Evangelicalism than intra-denominational believers.

Second, ethical sensibilities attuned to cosmopolitan, super-diverse, and pluralistic urban publics have shaped a more generous engagement with religious 'others'. While this can be manifest in tokenistic gestures that coexist with performed indifference, non-encounters, and liberal avoidance – all of which can conceal prejudice – it has more generally catalysed a wider changing consciousness towards the place of religion in the public realm. The propensity for postsecular rapprochement is both heightened and circumscribed by the context of the postcolonial city (Baker and Beaumont 2011a; Garbin and Strhan 2017). Moreover, the fast-changing plurality of belief, practice, and ethnicity has created opportunities for reflexive engagement and dialogical learning, which is finding expression in spaces of welfare, care, advocacy, and protest.

Accompanying these shifts has been a specific theological sea-change beyond religious fundamentalism which has galvanised ethical stances attuned to receptive generosity. Here we briefly illustrate characteristics of this theological shift by using the example of changing faith ethics within UK evangelicalism – a theological identity better known for its divisive postures towards the world. We have explored elsewhere the shifting landscapes of faith motivation that have led to what has been recognised as a rise in 'radical faith praxis' (Cloke *et al.* 2012), a movement from *propositional* modes of belief and ecclesial practice, towards more *participatory* and *performative* theologies that incorporate tradition and immanence in the form of theo-ethics (Cloke 2011a). Within UK evangelicalism, this has been embodied in a broader acceptance of a faith-imperative to act on issues of poverty and injustice, as well as recognising commonalities between theo-ethical values and the ethical values found in secular worldviews and other faith traditions. One of the notable currents shaping the capacities for receptive generosity has been ascension in theological discourse and practice of the notion of Missio Dei (Bosch 2011). Theologies of Missio Dei are characterised by two tenets: first, the idea that God's mission is wider than church activity, that the church is a participant rather exclusive workforce; and second, that the responsibility of the church is to find out what God is doing and 'join in'. The re-articulation of theologies of Missio Dei has permitted a release from fundamentalism on two grounds. First, it has urged doctrinal revision in some quarters of evangelicalism towards a more activist conception of mission. Missio Dei inspires an openness to put aside debates regarding 'exclusivist'/'inclusivist' concepts of salvation by acknowledging God works through all people, beyond the righteous silos of the 'saved'. Despite no clearcut translation across denominations, the amplified theological resonance of Missio Dei has led to more generous forms of interfaith partnership on issues of common concern. Second, Missio Dei inspires a critical self-reflexivity towards theological orthodoxies and the spatiality of faith practice. Moving beyond missional models that place expectations on others through invitations to 'our space' – be that a church service or the space of a 'service provider' – Missio Dei is embodied in radically receptive responses that seek co-presence in existing organic communal spaces wherever they are found. Barrett (2017) rightly cautions that this stance can potentially reinforce unilaterial missional stances that disguise colonial and paternal stances to others; and instead emphasises the importance of 'receiving' the Missio Dei in the self, from the other, and in our neighbourhoods. Theologies of Missio Dei have therefore encouraged a less self-identifying or authoritative approach to theological and sociological certainty, and instead recognise the importance of self-reflexivity in discerning the common intuitions across religious and secular actors motivated to 'do something' about injustice. This has been combined with a more poststructuralist articulation of religious belief that scholars argue usher in a new spirit of ecumenicism detached from dogmatic burden (Vattimo 2007). The turn towards performative forms of apologetics in which proclamation of the word is embedded in embodied performances of Christian caritas has led to explicit acknowledgement of how theo-ethics can cross over into partnerships with other groups and bodies working for the flourishing of the locality as a whole. While the bulk of faith-based social action still works within the parameters of church activity, these theological trends have given licence to more pluralistic faith-secular and interfaith collaborations. In such cases, we see a deliberate curation of faith spaces that seek to move beyond self-identification, assimilation, and exclusivity, instead attempting to create with others experimental thirdspaces where faith and spirituality can be acknowledged in pluralistic manner without the dangers of one-way moral freighting.

Elements of these subtle theological changes can be overlooked in blanket accounts of secularisation or understood in terms of anxieties surrounding the dilution of 'faith distinctiveness', but we argue for the significance of these movements beyond religious and secular fundamentalism that are being crystallised in spaces of care, welfare, and protest. These spaces are opportunities for rethinking theology, for example, how theologies of Missio Dei help galvanise new practices of receptive generosity across lines of social difference.

Postsecularity as active resistance

Postsecularity also seems to be mobilised in the circulation of values and practices that actively seek to deviate from the kinds of neoliberal subjectification that are manifest both in growing intolerance shown to the 'other' (immigrant, refugee, undeserving poor) and in the enculturation of neoliberal values in the ethical relation of self-to-self. Rapprochement between religious and secular ethics can therefore be understood as embodying a resistance to neoliberal politics of austerity. Neoliberal politics is underscored by ideologies of so-called fairness, supposedly 'just' rewards, and anti-dependency, all of which in practice are associated with a restrictive eligibility increasingly legitimatised by *moral calculations of behaviour* and *citizenship status*. Across a number of ethical and political arenas – for example, relating to homelessness, refugees, poverty, and welfare – faith groups have provided a sharp critique of the moral values under-pinning welfare as conceived and articulated through neoliberal economic metrics (Romanillos *et al.* 2012; Williams *et al.* 2012). While contestations between left-leaning religious theo-ethics and government are not entirely new (see Faith in the City 1985), what is significant in more recent faith-secular col-laborations and politicised expressions of public faith is the way in which the acceptability of religious voices in the public realm is contested. In the ideologically-charged context of austerity and Brexit, acquiescence seems to the principal fault-line on which faith groups are accommodated and enlisted by governments, whilst non-compliance with the parameters set by the state leads to the reprimanding of religious groups as 'out of place' in the public realm. In this way, secularity is used strategically as a performative device to defend ideo-logical agendas, as in, for example, strong efforts by David Cameron and the Conservative Party to parry the very public criticisms made by clergy that denounced the 'immorality' of welfare reform. For these reasons, left-leaning religious leaders and organisations are becoming prominent voices in a broader anti-austerity politics and defence of state welfare; although capacities for wider-level rapprochement between religious and secular ethics are undermined by right-leaning rapprochement of morally conservative religious voices in areas of gender, sexuality, and reproductive rights.

Nevertheless, the legitimation of dissident religious voices plays an important part in the counter-hegemonic 'resonance' machine in national public debate; and more locally can be shown to partner with others to work towards a broader urbanism of compassion, hope, and solidarity. Faith groups have been often acknowledged as 'gap-fillers' in welfare provision, particularly among groups deemed ineligible or undeserving of state support: the most obvious example is in the areas of asylum and single homelessness. Many FBOs deliberately resist partnership in joined-up modes of governance due to an unwillingness to com-promise on who they will and will not be allowed to serve. In so doing, the organisational space of FBOs can serve as a catalyst for secular and faith-motivated actors to recognise shared ethical commitments to radical expressions of universality and unconditionality; for example, churches operating informal networks of hospitality houses will sometimes 'hide' and financially support individuals and families who face deportation because of refused asylum status (see Howson 2011), and broad-based local secular and religious mobilisations will often seek to provide support for refugees and asylum seekers (as, for example, in the Bristol Hospitality Network).

The propensity for dissident religious voices has been heightened as a result of wider shifts between church, state, and society. In the move from Christendom to post-Christendom (Murray 2004), Christians now find themselves as 'one voice among many' and can no longer expect or benefit from the familiarity of institutional privilege and influence over public life. While some Christians lament the erosion of legislative powers with a sense of loss or even hawkishness, others celebrate throwing off the binds of established religion and its theologies of privilege that domesticated the good news to fit a particular charitable and nationalistic expression. This moment has opened new opportunities for the church to experiment with non-institutional spaces on the margins and embody a more radical, transnational movement of believers. It has also provided opportunities for a faith ethics of agape and caritas to find commonality with secular activists in a range of ethical, social, and political arenas. The multitude of dissident voices constitute a 'counter resonance machine' (Connolly 2008) challenging neoliberal calculations of value as well as the orthodoxies of institutional religion (see Cloke *et al.* 2016).

Affective life in late capitalism and the regime of desire produced through neoliberal subjectivities has also led to a bubbling up of self-reflexivity. This manifests itself as a critical awareness displayed towards the realities of cruel optimism, a disenchantment with the fantasies of the neoliberal myth of hard work and security, and a searching beyond the stasis of hopelessness. Crucially, self-reflexivity is also displayed in the admiration shown towards 'truth tellers' – those responding to the needs of people and speaking out against pernicious injustices that connect us all. Caution is obviously needed here to recognise ambivalence towards the 'truth teller' – a figure claimed across the political spectrum – and can risk sliding into fascist sensibilities and religious bigotry. Our claim, however, is that the subjectivities produced through the thirdspace of postsecularity relate strongly to this truth-telling role. Specifically, we refer to this 'truth-telling' act as a mode of spiritual contemplation that allows a discernment of and resistance against neoliberal enchantments (see Chapter 5).

Postsecularity and the affective politics of hope

A third expression of possibility within the subjectivities of postsecularity relates to the interconnection of affect and hope. Ideas about affect allow social science to recognise that which is invisible and unspeakable as nevertheless significant in the sourcing of emotion and experience. It follows that the hidden and ineffable qualities of faith, previously given short shrift by secular rationalists, are now not only being recognised for their affective capacity but also mapping more easily onto secular mysticism and faith in immanence. This space of convergence is especially relevant to ethical registers including those associated with hope, where currently blighted 'optimism' (that is a future pointed to by the present) is being transformed by a hopeful capacity to believe in what currently seems impossible. In particular, it is this affective politics of hope, we argue, which prompts collective energies of hopefulness that seem foolish and naïve in neoliberalised contexts dominated by affective atmospheres of fear and cruel optimism.

Caputo's (2016) poststructural theology of hope adds depth to this claim. He discusses two modulations of living: the risk-averse figure who seeks to preserve the present and resists what they cannot see coming – a form of reactionary conservative ethics; and the risk-taking figure who is willing to take a leap of faith by placing a belief, practice, tradition, institution, and the like in jeopardy in order to sustain its life and liveliness. In these terms, postsecularity represents the riskier relation to life. It embodies a hope in the impossible that is by no means an exclusively religious phenomenon; such hope is, says Caputo, contained in the name of God, but not contained by or confined to that name. In other words, hope is grounded in a kind of non-foundational theopoetics of the impossible. As explained in Chapter 2, a more poststructuralist articulation of religious belief has aided the practice of these theopoetics. As Christian faith has increasingly awoken to its new cultural reality after Christendom, so it has been drawn to faith-forms that emphasise the love and suffering, rather than the power and glory of God – a journey from the being of God to the story of God's being (Robbins 2007) especially with the poor, the hungry, and the outcast. Postmodern religious faith, then, looks to a more transcendent and less predetermined commitment to the essences of belief – according to Vattimo (2007) essences of agape and caritas that represent the scriptural limits to the desacralisation of society.

These theological currents have brought together important post-evangelical expressions of Christian faith, in which faithful relationships are more about agape and caritas than about truth, and hope is about releasing people to step into a destiny of non-dogmatic loving of others – neighbours and enemies alike. It follows that the opposite of a religious person is a loveless person (Caputo 2001) and that it makes sense to speak of the religious in people rather than of religious people *per se*. Thus, in Caputo's terms, religious and secular actors can both be lovers of the impossible, seeing beyond the foreseeable future to a 'God only knows what' future that is unforeseeable and where only the great passions of love, faith, and hope will see them through. Here we glimpse a remarkable possibility of postsecular subjectivities. Excessive love and perseverance express an obstinate hope for transformation; a venture into the hyper-real carrying an impossible belief in the seemingly impossible. Whether in the abolition of slavery, or the fight against racial segregation and for racial equality, or (and here insert your currently unforeseeable hopefulness), people have really needed to believe such things were possible before they could possibly happen. Faith, then, is not just a motivation, but a capacity to shape affective engagement with the world. It is embodied in religious and non-religious ways, embracing an ethical subjectivity involving the turning of the self towards the transcendent and the immanent. It requires what Habermas (2006c, 30) refers to as the 'costly commitment' of acting on behalf of others, an ethical disposition that late capitalist liberal democracies have struggled to engineer because of the embeddedness of neoliberal logics in personal, aesthetic, and affective registers. However, religious articulations of faith, hope, and love seem to have a role to play in shaping a more progressive politics of hope. As Critchley (2012) argues, fidelity to these ethical capacities – whether religious or not – offers the hope in the seemingly impossible; a prying open of ethical and political spaces that rework and resist callous neoliberal subjectivities.

Postsecularity as post-disenchantment: new ethics, values, expressions

The fallout of neoliberal excess has presented a deeply fractured context for the cultivation of subjectivities for postsecularity. While the idea of re-enchantment has to be recognised as complex and multifaceted, we want to argue that this crisis of secular – and religious – consciousness enables new spaces and subjectivities of spiritual disobedience in which deep-seated hierarchies, predispositions, and affective capacities can be reworked. Having to respond 'in the meantime' to the mean times cultivates a terrain in which forms of re-enchantment can emerge; moments of reflexive postsecularity can bubble up almost regardless of the setting, simply because alternative ethical predispositions and affective politics enable a mysterious sense in which the subjectivities of desire become open to remodelling. It is possible, therefore, to imagine a reshaping of desire in these terrains, away from those values inflicted by self-interested capitalism, and towards values fed by a counter-cultural, and sometimes theological ethics that confront the prioritisation of wealth and self-pleasure with an emergent affective capacity for hopefulness and healing, hospitality and generosity, justice and equality.

Postsecularity at its core connects to the *politics of ethics* – how ethical relations are organised, distributed, negotiated, and performed. Here, we suggest Rose's (2017) notion of 'post-disenchantment' is a helpful device to foreground the co-constitutive forces of disenchantment and enchantment at work in late capitalist societies. We argue that subjectivities of postsecularity help to co-create a critical awareness of the enchantments of consumerism, the false promise of hard work and meritocracy. These subjectivities can also prompt a disenchantment with 'traditional' religious forms and practices and a search for more enchanting faith expressions. Rose deploys the notion of 'post-disenchantment' in order to challenge the idea that we live in a 'disenchanted' world where the appropriate solution is a process of 're-enchantment'. Rather, the case for post-disenchantment

> recognises not only that we can no longer evade the persistence of magical thinking and practices, but also that the entangling networks which constitute contemporary capitalism function as a system of technological re-enchantment, a secular reiteration of the kinds of structures of power and dominion which characterised the enchanted universe of classical Christian thought.
>
> (p. 243)

Applying this understanding to the nexus of emotion, belief, and ethics focusses our attention not only on the ways in which late capitalism operates through enchantment (also see Bennett 2001; Hilgers 2014; Bartolini *et al.* 2016), but also on the resonances and dissonances between late capitalist enchantments and religious enchantment which currently shape much of religious belief, ethics, and expression.

Our argument is that postsecularity helps attune critical sensibilities towards unlearning the enchantments of right-leaning religion and late capitalism. As the existential burden of late capitalism increasingly becomes a visceral experience, and the myths that inspire cruel optimism (Berlant 2011) are increasingly revealed for what they are, there will be a bifurcating effect on religion itself. On the one hand, the appeal of religious fundamentalism and its apparent offer of emotional solace in theological certainty and 'otherworldly' enchantment can be shored up; on the other hand, the existential disconnect between promise and reality, and between the myths of certain religious and neoliberal orthodoxies and the shacking up of injustice and disconnection from others, can create anxieties of bad faith and normalised cognitive dissonance. Here, there are opportunities to reformulate practices and expressions of religious and non-religious belief so as to avoid amplifying resonances that embolden capitalist enchantments (Connolly 2008). It is in this release from the ways that neoliberalism shapes values, and the ways that right-leaning religious enchantments have shaped ethical sensibilities, that new forms of values and ethics reveal themselves. Even in the darkness that this assessment of reality often brings on, the immense interdependence of the universe can still produce the unexpected. Grace can emerge. A politically and ethically charged re-framing of enchantment – albeit often fragile and ephemeral – can prompt practices that contradict, and act as a bulwark against, the 'bad faith' of capitalist enchantment.

This search for authenticity and re-enchantment is emerging from both within religious traditions as well as secular ones, as many believers are looking for new and more challenging ways to live out religious precepts in the world, subverting or directly challenging what many see as either over-prescriptive or over quietist and comfortable religious subcultures (Bielo 2013; Baker 2017a; Thomas 2013). While it is easy to characterise post-disenchantment in terms of politically right- or left-leaning religious trajectories, we are interested here in the less obviously partisan mysticisms of theological enchantment that contribute to new psycho-geographies of performance and praxis drawing on 'deep enchantment', which can generate more assured postures of collaboration that need not fixate on doctrinal or ideological closure. The enchantment in subjectivities of postsecularity, then, may most usefully be considered as a/theistic rather than religious, formed from an upwelling spirituality of reflexive personal desire and re-sensitised ethical engagement that resonates with a desire to join with others (secular or religious) who share similar instincts towards post-disenchantment; whose trust and belief in new forms of hopefulness inspires them to act.

3.4 Three examples of postsecularity

To end this chapter, we draw on three examples to illustrate empirically the ways in which postsecularity emerges in situ and serves as a catalyst in the release of new ethical capacities and subjectivities. These examples emerge out of the three possibilities generated by postsecularity identified in the previous section, namely: as a movement beyond religious and secular fundamentalisms; as a source of intentional and active resistance against neo-liberal subjectification; and as the releasing of new ethics, values, and expressions as an embodiment of a new post-disenchantment. In each example, these characteristics of postsecularity are shown to emerge organically rather than being intentionally designed or curated. Our purpose here is to focus attention on the defining characteristics of postsecularity that come to the surface as a result of practices of receptive generosity, rapprochement, and hopeful enchantment.

Re-scripting the city through receptive generosity: the Pauluskerk

As discussed in Chapter 1, receptive generosity refers to an ethical stance towards others that is characterised by a deliberate forgoing of privilege, position, and agenda in order to pursue more reciprocal relationships. It is in mutual self-reflexivity – a process of making space with the self for the other – that ethical practices of in-commonness can be nurtured. Put differently, it is the willingness to be generously receptive to the being and voices of others and a wish to provide something of value. Here we want to illustrate characteristics of this receptive generosity through the example of radical gestures of hospitality during the 1980s–1990s towards a highly stigmatised population of drug users in Rotterdam, the Netherlands.

Perspectives on Christian involvement in drug and alcohol treatment have usually focussed on narratives of control, rehabilitation, and religious indoctrination (O'Neil 2014; Williams 2017). In the USA, for instance, court-ordered referrals to evangelical rehabilitation centres are said to constitute the biopolitical underbelly of the 'war on drugs' and the carceral-industrial complex (Rodriguez 2010; Bourgois and Hart 2010). While religious ethics certainly shape the spatial politics of intoxicants and addiction in cities, it is important to note their divergent expressions. In Rotterdam and Sydney, for instance, radical faith-inspired notions of sanctuary are deployed as a subversive theo-ethic that lays claim over space beyond jurisdiction of state law and usurps stigmatising socio-spatial discourses of the 'addict' (Williams 2013; Prior and Crofts 2015). We illustrate this with the example of the Pauluskerk (St Paul's Church) located in the heart of a deprived district of central Rotterdam, and which served as a shelter and day centre for people experiencing various forms of homelessness and substance misuse as well as for undocumented migrants. In the 1980s the Pauluskerk pioneered harm-reduction services in the city providing needle exchange, advice, and counselling services and, more controversially at the time, supervised rooms for heroin and cocaine users to take their drugs in a safe and sanitary environment. The initiative started after the Dutch Reformed Church gave Reverend Hans Visser carte blanche to develop a vision of an inclusive church amongst the homeless, 'addicted', and socially excluded populations in the city. Visser, a Dutch missionary who had spent ten years in Indonesia, returned with a theo-ethical vision driven by Jesus' parable of the sheep and the goats in Matthew 25: how we love the 'least of these' – which he read to be homeless people, drug users, and undocumented immigrants – is a reflection of love for Jesus. The Pauluskerk seemed to challenge the moral dogmatism of religious and secular people in Dutch society. We want to draw out some aspects of the Pauluskerk that seem to embody different aspects of postsecularity, and as they relate to receptive generosity.

First, in making space for the other within the self – physically, financially, and socially – the Pauluskerk embodied the ethos of receptive generosity exemplified in Coles's (1997) notion of postsecular caritas. Opening its doors as a place where people could meet and find comfort in the form of a bed, a meal, a conversation, or even a drug in a safe and nurturing environment, the Pauluskerk embodied a non-judgemental ethos that was receptive of 'otherness' – not demanding the 'other' should change their behaviour or identity in order to come and experience community. The Pauluskerk's ethos of acceptance derived from the theological notion of Christ's incarnation – becoming fully flesh and fully engaged in a messy world. In welcoming 'the other', the Pauluskerk challenged the dominant stigmatisation of substance misuse presented in dogmatic 'tough love' mentalities: the belief that drug users who fall through the welfare safety net have themselves to blame and do not deserve help: that it is better for them to hit 'rock bottom' sooner to recognise the need to change their ways. The Pauluskerk's radical understanding of Christian caritas and God's unconditional love for all people leads to an act of hospitality that challenges moral notions of (un)deservingness.

Second, the Pauluskerk's way-of-working, creating safe spaces for the most marginalised to become part of the Pauluskerk community and hearing the stories of addicted marginality, can be considered to be generative of new theologies and politics of hospitality and generosity. While some questioned the morality basis of Visser's approach on legal and theological grounds, the radical ethos of engagement premised on acceptance and listening spaces to be in-common with others can present a challenge to orthodox medical and theological thinking on 'addiction'. It serves as a liminal space where actors come to question the veracity of the dominant addiction-as-pathology paradigm (the biomedical 'brain disease' model), which have helped shape, and have been shaped by, theological views of addiction as moral failure. Liminal spaces of encounter, as witnessed in the Pauluskerk, potentially can solicit experiences in which we welcome the figure of the drug user not as someone to be 'fixed', but rather as a prophetic gift – one which speaks about the narcotic experience of modernity itself (Derrida 1993 in Smith 2016), its intoxications, escapism, and phantasmagorias. In this way, the figure of the drug user can be read as an unwitting prophet (Dunnington 2011) who challenges the expulsion logic within the self, church, and society. Through listening to marginal voices and experiences, certain theological proliferating practices of 'tough love' and 'rock bottom' might become problematised, and informed by a more embodied and intuitive understanding of the structural and systemic forces that serve to produce and perpetuate harm for people who use drugs: repressive drug laws; socio-spatial inequalities; the stigmatisation of users, especially with regard to gender and race; treatment modalities characterised by illiberal intervention and punitive discipline (see Smith 2016, 211–216). Harm reduction philosophy was initially an act of civil disobedience by users and health professionals who sought to alter the material and social conditions of addiction and sharpen critical 'political ana-lysis of "risk" and "harm" as by-products of social, economic, racial or political inequality' (Roe 2005, 245 cited in Smith 2016, 215). It is possible therefore for mundane spaces of being in-common and listening to voices of marginalised 'others' to be channels through which we call into question the ways theological constructions of addiction – directly or inadvertently – become enrolled in exclu-sionary and harmful practices, and maybe develop more subversive or anarchic theo-ethics that explicitly seek to oppose and challenge the regimes that produce harm for people who use illicit substances (Smith 2016).

Despite an uneasy relationship with local community and state authorities, the Pauluskerk developed good working relationships with public health and social services, which provided a doctor and three mental health nurses, as well as medication and supplies such as needles and condoms. Such a rapprochement between a radical faith-inspired project and secular authorities, allowed people access to a range of support services who might be not be able to make use of existing services because of lack of insurance or undocumented status. However, shortly after Visser's retirement, the Pauluskerk scaled back its activities as a new purpose-built centre was erected on the site. This followed several years of often tense negotiation between municipality and the church (For more informa-tion on the political and organisational context, see Davelaar *et al.* 2013; Wil-liams 2013).

Nevertheless, the activities initiated by Pauluskerk helped catalyse wider reformulations of state responsibility for marginalised drug users in the city. By embodying an alternative gesture of welcome, non-judgementalism and accept-ance, the Pauluskerk provided an iconic space that challenged the prevailing dis-course and the logic of possibility surrounding drug treatment intervention. This example shows how receptive generosity to difference became a radically sub-versive moment in Rotterdam. Set against a hostile backdrop of stigmatisation, the Pauluskerk represented elements of a 'temporary autonomous zone' for mar-ginalised groups in society, a place free from the confines of law and social norms, creating space capable of re-imagining social relations and identity. The Pauluskerk's central location in the city challenged the scripting of public space and the rights to the city more directly. Its ethos of unconditionality injected hope in overpowering situations, giving glimpses of what a new world might look like; inviting participation in a countercultural vision of a different order: one that makes a spectacle of moral notions of 'deservedness'. The Pauluskerk represented the potential of an FBO to paint prophetically alternative pictures of the city (and the church), through the denouncement of the socially excluding practices of containment, criminalisation, and 'othering' that work to stigmatise those struggling with addiction and substance misuse.

Subjectivities and practices of rapprochement based on the jungle camps

In this section we explore the complex nuances associated with new subjectivities and tactics of rapprochement and postsecular partnership with respect to the volunteer humanitarianism response to the Syrian refugee crisis in 2015–2016. These expressions of rapprochement found their most proximate and visceral expression for Northern Europe in the so-called Jungle, or squatter camps that emerged around the port of Calais and more latterly, Dunkirk. The origins of the Jungle lie back as far as 1999, when at the request of the French government, the French Red Cross opened a humanitarian warehouse at Sangatte to provide short-stay accommodation for refugees on their way to the UK. However, in 2002, the centre was closed after pressure from the British government who regarded it as a 'pull factor' for further undocumented migrants and refugees. The subsequent problem of refugees sleeping rough in the Calais area whilst attempting to cross illegally into the UK became a vexed and long-running issue. By the start of 2015, the numbers of those squatting in these impromptu sites was estimated at 2,000. In order to minimise disruption to sites closest to the travel terminals, the local authorities attempted to encourage relocation to a single spot – a dumping ground on the outskirts of Calais, called the Jungle (King 2016). However, due to the new humanitarian crises prompted in the main by fresh onslaughts in the Syrian war but perpetuated by ongoing conflicts in countries such as Afghanistan, Eritrea, Somalia, and Kurdish areas, the population of the Jungle rose steeply to over 9,000 people by August 2016, including nearly 700 unaccompanied minors, the youngest of whom was eight years old (Sandri 2018). This exponential growth in vulnerable refugees was exacerbated by obvious healthcare and sanitation issues, but also the immiseration and fear caused by the entry of trafficking gangs into the camps and their attempts to extort money from different national groups and factions.

But before going into the details of the new and intricate webs of solidarity and ethical concern that emerged in response to this deliberate state-sponsored and violent neglect, it is necessary to define the theoretical lens through which this event is being framed. As Kim Rygiel (2011) has pointed out, the academic literature that has built up around the issue of displaced migrants since 2008 has reflected two competing theoretical tropes. The first tradition is essentially (and not surprisingly) predicated on narratives of loss and emptiness. This tradition, influenced strongly by the work of Giorgio Agamben, portrays those incarcerated in refugee and other camps as citizens whose rights have been stripped off them by the exceptionalism of the state, and who are condemned to 'bare life' (1998). In this line of argument, the power of the state, as Sovereign, lies in the 'ban' or the exceptional power to shun a person from a political community. Agamben coins the concept of *homo sacer* – namely 'the one who may be killed and not yet sacrificed' (1998, 7) to further describe the reality of life and subjectivity of those systematically excluded from the political community.

However, some criticise Agamben's emphasis on state exceptionality since it ignores the specific and historical material reality of camps and reduces them to abstract and depoliticised spaces (see, for example, Walters 2008; Isin and Rygiel 2007). Thus, Rygiel identifies a counter trajectory of theoretical representation which stresses the 'autonomous migration' (2011, 3) narrative of citizenship. This tradition highlights 'migrant's agency' whereby their 'mobility' is a resource that allows them to potentially confound and destabilise 'the distributions and markings of social power' (ibid.). From an autonomous migration perspective therefore, border controls are interpreted as 'following or reacting to migrant's own pathways of movement rather than preceding them' (2011, 4). With regard to notions of citizenship, Rygiel makes the important point that despite authorities' attempts to create the meaning of the camp as a space for 'bare life', the disruptive agency of the migrant and refugee always presents a dangerous alternative. Thus, these spaces can also potentially be sources of social and political autonomy from which a politicised citizenship dares to assert itself. Sandri also runs alongside the 'insurgent citizenship' lens the idea of the camp as an assemblage (Ong 2007) which includes both material and human actors, and therefore the volunteers as well as the refugees themselves. The fluid and relationally-based interpretation offered by Assemblage and Actor Network Theory (Latour 2005; Blok and Farias 2016) allows her to pull together key tropes and ideas that are highly pertinent to our reading of the tactics and subjectivities of rapprochement wrought by postsecularity.

The first dimension of rapprochement concerns Sandri's analysis of volunteer humanitarianism as an affective and ethical response to the processes of neo-liberalism that *blurs the boundaries between voluntarism and activism*. Whilst it would be easy to dismiss volunteer humanitarianism as an unwitting response to the intense trajectory of the neo-liberal project (namely the outsourcing of civil society to private individuals), Sandri's assemblage analysis points to a more complex picture. On the one hand, she argues, the strong activist networks generated as a response to the human rights abuses in the camps does not collude with the extended governmentality of the neo-liberal state (Ferguson and Gupta 2002) because the volunteers are actually inserted within the meaning-making of the potentially insurgent space of the camp, thereby 'challenging the state about the violent border practices and the asylum regime' (Sandri 2018, 76). What is clear from Sandri's research as a participant observer with a large voluntary humanitarian network (20 core members and 500 casual volunteers) is the presence of alternative affects and values which are driving the creation of this space. In what has been described as the 'care-justice transition' (Williams *et al.* 2016, 2307), Sandri's participants working in grassroot organisations providing aid to refugees predominantly expressed their initial motivation as primarily humanitarian not political. The stance of Lotus (the pseudonym of this humanitarian network) differed substantially from other organisations in Calais that were explicitly driven in their motivation by border contestation and solidarity activism. Indeed, Sandri notes,

> The arrival of a multitude of volunteer humanitarians in the summer of 2015 was met with hesitation by activist groups which had already gravitated around the Jungle due to political convictions, such as CMS (Calais Migrant Solidarity, a division of the activist group No Borders).
>
> (p. 75)

However, after experiencing the harrowing conditions of the Jungle, many volunteers became activists for the plight of the Jungle immigrants when they returned to the UK. Lotus found itself positioned at the forefront of activism in the UK, contesting the violence of border control and putting pressure on the British government to improve living standards in the Jungle, address French police brutality, and to recognise its global responsibility to welcome more refugees (ibid.). For some volunteers, like Hannah, the experience represented a conscientisation:

> At the beginning I thought I was going on purely humanitarian grounds. But it is very difficult not to realise that this type of caring becomes politically motivated. A human right is that people are safe and not abused, especially when they are so vulnerable. So, this side of things made me feel more politically motivated.
>
> (cited in Sandri 2018, 75)

One of the hallmarks of geographies of postsecularity, we are suggesting, is the blurring of traditional lines of demarcation – and one of these lines is the blurring, or perhaps more accurately, the continuum between activism and humanitarianism impulses. However, Sandri makes it clear that this humanitarian impulse is never totally subsumed within a narrowly defined activist stance. While Lotus volunteers refused to cooperate with state authorities and became enrolled in bureaucratic mechanisms that might jeopardise an individual's claim to asylum, their central motivation was driven by a humanitarian concern to alleviate suffering and they were wary that more politicised activity inside the camp might attract police intervention (see Sandri 2018, 75–76). Yet, back in the UK, the events, experiences, and relationships in the Jungle spilled over in the lives of volunteers, many of whom joined political parties, became refugee advocates in their social networks, organised awareness campaigns and demonstrations that challenged the pernicious injustice of border control and the government's deliberate *in*action that exacerbated the suffering of refugees (Sandri 2018, 76; also see Davies *et al.* 2017).

A second dimension of postsecular rapprochement emerging from this case study is a form of coming together and partnership that is based on *a shared vulnerability and a common search for meaning and wellbeing.* The camp generated an affective sociality between disparate sets of personalities and identities with the capacity for new affinities and subjectivities to be made. Sandri, for example, draws on Malkki (2015) to suggest a distinct social and affective atmosphere is produced through the in-common pursuit of humanitarian justice, particularly through giving up regular weekends, energy, and waged labour. Volunteer subjectivities show signs of what Caputo calls 'hope in the impossible' and the ontological courage to be unhinged, to seek the promise of justice despite overwhelming odds (see Caputo 2016, 9). Interestingly, for some volunteers, the pull towards these spaces was linked to disenchantment with late-capitalist rhythms of the everyday, and volunteering was seen to offer 'protection against loneliness, social isolation and asocial time' (Malkki 2015, 134 cited in Sandri 2018, 76). These social and almost existential needs were met in new affinities of response expressed in the voluntary humanitarianism of the camp that transgressed the usual barriers. Volunteers with contrasting backgrounds and ideological motivations were drawn together into networks of car-sharing and organising aid convoys. However, these affective ties between volunteers also crossed over in the forging of friendships based on emotional and personal connection between volunteers and refugees, and these connections gave volunteers an even stronger sense of purpose, as they exercised responsibility towards people who they counted as friends living in the camp.

A third dimension of rapprochement amplified in first-hand blog accounts of the volunteer humanitarianism that took place in the Jungle camps *involves the search for belonging and deeper resonance with humanitarian issues*. Much of the first hand written and pictorial accounts of life in both the Calais and Dunkirk camps are still available in on-line archives and curated collections, for example, at www.calaidipedia.co.uk. Access to this first-hand narrative is particularly useful for identifying other dimensions of relational postsecularity that we have identified above.

For example, there is the impulse for attaining some sort of deeper resonance and meaning making that joining affinity action groups such as those that emerged in response to the Jungle represents. It is the searching after an 'adventure' or 'experience'. The following blog captures this dynamic very well:

> I am not a political person, nor am I activist. I work three jobs in the fashion industry. I love makeup and fashion and I am obsessed with all things celebrity. I am 26 years old and I have never been outside of Europe, except for a brief trip to the States a few years ago. I saw the photos, like most people, of little Aylan's body washed up on the beach, and as I had some free time I just thought 'why not go over and help?' I had seen photos from other volunteers online, of them doing art workshops and watching movies with the kids, just making them smile in general – I wanted to do that! I racked my brains as to how I could get out of going, but in the end decided to just do it. I kept telling myself, and other people, that it would be an 'adventure' and 'an experience'.

> (Shortall 2015, para 1)

The blog archives capture the deep changes in cognitive, ethical, and sometimes spiritual perspective that is generated by these affectivities rooted in hospitality and vulnerability. In this series of extracts written for the Huffington Post, we see how the disruption of power dynamics enacted by a reversal of roles of donor and recipient in the camp (a perpetual theme) often causes a crisis of identity, knowledge, and understanding in the experience of the donor. Most often this power reversal is expressed in the sharing of tea and food, these resources being offered by the camp citizens to the volunteers from their very meagre resources. Santosh Carvalho, a volunteer, recalls one such encounter with an Afghan soldier who had survived being shot in an execution attempt by the Taliban:

> Sharing tea with Shapoor is such a humbling experience. His hospitality is overwhelming despite having nothing. He keeps offering me every last bit of food he has. In his culture, the guest is looked after first and foremost. What does that say about us as a society, where we have everything but aren't prepared to share anything?
>
> (Carvalho 2017, para 16)

As Santosh reflects on the deeper cultural significance of this experience, he goes into the camp school to teach English lessons. Aware of the emotional and physical stress this work places on him, he once again finds an unexpected uplift and hope from the encounter. But this time the change wrought upon him is perhaps more spiritual than intellectual:

> The highlight of teaching today was being told by a Sudanese student that he will remember my teaching for the rest of his life. I was almost moved to tears. I cannot believe that I have affected a life in such a short space of time. I am becoming increasingly spiritual/religious here. I will pray for each and every person I have met here.
>
> (Carvalho 2017, para 24)

However, this energetic and creative weaving of personal trajectories between voluntary, institutional, and activist hubs needs to be put into a broader context. Relationships and partnerships generated by rapprochement can create progressive synergies, but also leave some other important issues unaddressed. On the positive side, volunteer humanitarianism had a direct and necessary impact. Shelters were built, clothes were donated, and hot food prepared, all of which made a beneficial impact on the lives of the refugees. However, there were limitations and downsides to these informal interventions that other institutions or agencies would have needed to address:

> Without any general management of the personnel, volunteers were not accountable, and they could constitute a liability ... There were cases of volunteers being banned from working in the camp because of their lack of

professionalism or because they were taking advantage of the refugees. Volunteers were able to help at crisis points but struggled to deal with more complex situations that arose, such as trafficking, exploitation and violence ... The lack of institutional aid was worrying not only because there was a deficit of basic services, but also because it put refugees and volunteers at risk.

(Sandri 2018, 74)

The rich data emerging from the Jungle case study shows, to use Sandri's term, a 'constellation' of new partnerships and affiliations that arise from geographies of postsecularity. These forms of rapprochement are often highly creative, effective, rely on deep levels of trust, communication, and openness, and have life-long impacts upon those who choose to participate in them. However, they are also prone to people's good will and ethical impulses being exploited by others, which also confers an actual vulnerability and fragility borne from a lack of safety and accountability on all those caught up in these spaces of rapprochement.

Reshaping desire? Subjectivities of postsecular enchantment in Christchurch, New Zealand

In 2010 and 2011, the New Zealand city of Christchurch was devastated by two major earthquakes and the thousands of aftershocks that followed them. The 2011 quake resulted in 185 fatalities, thousands of people injured, and damage beyond repair to significant areas of the central city and suburbs (see Adamson 2017; Pickles 2016; Roome 2012). Seventy per cent of the buildings in downtown Christchurch had to be demolished, and several suburbs were evacuated and cleared. No residents of the city were left untouched by the impact of this devastation; for many the earthquakes necessitated the social displacement of forced migration, and for others there was a negative emotional atmosphere, often made worse by a protracted battle to obtain equitable insurance settlements for damaged properties (Miles 2016). In addition, the earthquakes deconstructed the existing ideological and mythmaking narratives of the city itself. Prior to the quakes, two dominant narratives prevailed. First, the city traded on its history of quaint Englishness (Cupples and Glynn 2009); not only had pre-colonial traces been largely erased, but a dominant Anglophilia pervaded the built environment and the trees, parks, and gardens that characterised the self-tagged 'garden city'. Second, New Zealand has been recognised as a laboratory of neoliberalisation (Rashbrooke 2013), with long-term adherence to monetarist economic policy, reduction of state intervention and subsidy, and privatisation of state assets. This national context of neoliberal experimentation shaped the narrative of urban development in Christchurch, where privatisation, deregulation, and state shrinkage dominated, and a largely permissive regime of planning and regulation gave developers considerable freedom to shape the urban environment as they wished (Hayward 2012).

These narratives of the city were ruptured alongside its built infrastructure by the earthquakes. However, as Solnit (2009) has emphasised, disasters often seem to open up opportunities for political and/or socio-cultural innovation. Recent research (see Cloke and Conradson 2018; Cloke and Dickinson 2019) has charted how the previous neoliberal and neo-colonial shaping of desire in the city has been challenged by the activities of transitional organisations who have been promoting community-minded expressions of the city through theatrical and artistic interventions and experiments which have made temporary use of some of the clearance sites in the city centre and elsewhere. In broad terms, these activities have been built upon a strong sense of disenchantment with the top-down urban neoliberalism of blueprinted development in concrete and glass. Instead, their use of temporary materials and sites and their championing of grassroots action have evoked different senses of desire, based on attempts to seek out forms of re-enchantment built on particular ethical platforms enabled by the rupture of the earthquakes. Specifically, transitional organisations have emphasised the importance of experimentation (with an inherent forgiveness for the mistakes that inevitably occur with some experiments), the civic rights associated with the common good and the practice of in-commonness (with an associated rejection of political-economic and socio-cultural elitism), and the performance of aesthetic connection (with an accompanying emphasis on people's capacity to discern meaning without being more formally propagandised). In so doing, it can be argued, they have acted against previous disenchantment and started to find new platforms of desire on which to found a re-enchantment of the city.

How then, might this example be seen to illustrate aspects of the bubbling up of subjectivities which are connected to postsecularity? To start with, New Zealand has a rich tradition of integration of religion and society; although church and state have always been held separate, the welfare state introduced by the first Labour government in 1935 was regarded as a system of practical Christian compassion (Franks and McAloon 2016). In line with other contexts, religious observance in New Zealand has declined since the 1960s, but the 2013 census indicated that around half of the population remained affiliated with Christian religion (Lineham 2017). Recent research by Sibley and Bulbulia (2012) even suggested that levels of religious faith actually increased amongst earthquake-affected people in Christchurch. This socialised context of religion could be set aside as simply part of a more widely recognised collaboration between conservative religious values and neoliberalism (see Hackworth 2012), in which the practical impact of societal religiosity is to reinforce the privatisation, individualisation, and deregulation of political activity to the detriment of socially and economically disadvantaged people. However, there is evidence in post-earthquake Christchurch that a rather different form of collaboration has been achieved across secular and religious boundaries in the city that has contributed to a new 'experiencescape' (Johansson and Kociatliewicz 2011) in which re-enchantment is emerging through a series of interventions that suggest a rather different 'soul' of the city (Fuste-Forne 2017).

A very significant element of this new collaboration has been the capacity of many Christian churches in Christchurch to move away from a perspective that identifies disaster-response as an opportunity for evangelism and proselytisation, or as a platform for moralistic connections between the earthquakes and divine outrage. While these tendencies were not completely absent, Brogt *et al.* (2015) suggest that churches made an important contribution to the resilience of the city precisely because they performed less fundamentalist, and more socially just and civic-oriented roles. The long-term pursuit of Christian social action and the short-term exigencies of earthquake relief combined to enable churches both to challenge the values underpinning neoliberalised registers of desire, and to help pave the way for alternative assemblages of compassion, acceptance, and experimentation. The new experiencescape of Christchurch, then, incorporated new kinds of subjectivities involving postsecularity, as religious and secular individuals and organisations worked together to re-energise the transcendental role of place to connect up particular activities and events with an emergent local identity.

The story of Christian responses to the Christchurch earthquakes has been narrated in detail by Parsons (2014), but for the purposes of this illustration of how such responses contributed to the process of re-enchanting the identity of the city, we can point to four significant performativities that reflect how subjectivities of postsecularity bubbled up amongst the shattered illusions of security in property and possessions. First, in the immediate post-quake period, churches acted as a crucial cog in the civic response to peoples' needs. They provided facilities for shelter and for the distribution of emergency resources, they were an important source of volunteers for wider emergency tasks, and they represented a largely trusted means of access to hard-hit local communities, for example, by house-to-house door-knocking in the worst affected suburbs to check on individual need. Second, churches exercised socially sensitive pastoral care in the wider community, not least through chaplaincy roles in frontline organisations such as the police, the fire services, and hospitals where to be present alongside staff (whether religious or not) was to perform crucial 'care for the carers' who themselves were traumatised by crisis and grief. This emergency involvement continued into wider psychosocial support through organisations such as the Petersgate Counselling and Education Centre, an agency supported by mainline Christian churches but offering a professional secular counselling service. Petersgate brought together faith-based and secular people and resources into a partnership of postsecularity in order to provide an accessible and socially trusted centre for trauma counselling, and thus contributed to the longer-term coping with emotional and affective responses to the earthquakes at community and city scales. Third, in a series of civic memorial events held to remember the city's losses, the multiple contributions made to post-quake well-being, and the hopes for the city's future, religious involvement in the expression of lament over recent past events and of the significance of neighbourliness, graciousness, and open-handedness to the overall spirit and soul of the city, helped to establish a tone for a re-oriententation of desire in Christchurch.

Fourth, faith-oriented people and organisations have also played a part in the venting of new transitional urbanisms and ethics from the ruptures caused by the earthquakes. Despite the top-down plans for a neoliberalised blueprint for the rebuilding of the city, Christchurch has seen a minor explosion of alternative tactical urbanism, involving the deployment of small-scale transitional interventions to suggest alternative bases for longer-term change. This transitional urbanism in the city has been detailed elsewhere (see Bennett *et al.* 2014) and recent research (Cloke and Conradson 2018; Cloke and Dickinson 2019) has suggested how the performance of urban experimentation involved has punched far above its weight in terms of introducing new grassroots participation and new ethical codes of in-commonness into the experiencescape of Christchurch. The centre of the city is now peppered with murals, artwork, micro-gardens events, performances, installations and even temporary buildings that attest to a rejection of the previous unevenly distributed and neoliberalised neo-colonialism, and a desire for greater senses of community participation and involvement in the unfolding of the post-quake city. The key organisations behind these initiatives – Gap Filler, Greening the Rubble, Life in Vacant Spaces (LIVS) and so on – have emerged from communities associated with radical arts, environmentalism, and architecture in the city, and have no obvious allegiances with faith-based organisations and individuals. However, their ethical codes emphasise the common good, civic rights and an openness of participation, and their practices of grassroots experimentation include an open-handedness across secular-religious divides. Ryan Reynolds (a key figure in Gap Filler and LIVS) speaks of a 'gap' that has provided 'a space that allowed people to project their own desires, at least some of which spoke to … what Christchurch could be – less stuffy and conservative, more inclusive and collaborative, open to trying new things' (Reynolds 2014, 169). These new transitional urbanisms, then are open to, and provide space for, the involvement of different kinds of postsecularities – in particular those that move away from stuffy conservatism and desire new forms of inclusivity. Two examples reflect this openness.

At least some of the street art that has emerged in Christchurch city centre embraces religious and faith-related themes of lament and hopefulness. For example, one of the key memorial artworks in Christchurch is Peter Majendie's '185 Empty Chairs'. A Christian artist who is the driving force of the Side Door Arts Trust, Majendie began to accumulate a collection of individualised chairs, each one representing someone who had died in the 2011 earthquake. Each of the chairs was hand-painted white, and when displayed together they signified both the individuality and collectivity of the tragedy. The emotional potency of the memorial was evident not only in the Visitor Book commentaries, but also in the way the installation became a profound and valued expression of civic memorialisation; although it was only intended to last two weeks, 185 Empty Chairs has so far been retained as a marker of the spirit of the city in its current phase. Equally, and on a somewhat grander scale, the Transitional Cathedral also serves as a marker of postsecularity in amongst the transitionalism of the city. Following the severe damage to the original gothic-style cathedral, the Anglican

Diocese constructed a transitional cathedral building designed by Shigeru Ban which became the first non-commercial structure to be built in the city centre after the earthquakes. The transitional cathedral quickly became the city's most important building and an iconic symbol of experimental urbanism in the city; the predominant building materials were a series of massive cardboard tube rafters that supported its A-frame design (although wood and polycarbonate were also used, and the base for the frame consisted of metal shipping containers). This 'cardboard cathedral' also incorporated recycled wood and glass from damaged city buildings, thus encapsulating memory into the DNA of a structure which was the antithesis of 'stuffy and conservative'. The building is used for religious services, but it has also become a key venue for the wider non-church community as a place for a broad range of performances, meals, and gatherings, thus enabling further opportunities for the emergence of postsecular subjectivities. Interestingly, legal battles have ensued over the question of whether the original cathedral should be reconstructed (in recognition of its contribution to civic historicity) or replaced by a more serviceable structure (as desired by the Diocese). The decision to repair rather than replace demonstrates the continuing power of conservatism in the city.

These various performativities – from emergency relief to transitional structures – are significant because they reflect, at least in part, an ethical re-orientation of neoliberalised desire. In the aftermath of the earthquakes, there have been important attempts to open out spaces of in-commonness, grassroots participation, and experimentation, each in their own way seeking to re-enchant common life in the city by combatting individualism, building community, and collaborating in joyful and hopeful projects. In other words, people in Christchurch have been learning to stay with the trouble (Haraway, 2016) – experimenting with a new public consciousness of how better to live together in their portion of the damaged earth. The outpouring of receptive generosity, and the willingness to partner either explicitly or intuitively across religious-secular divides, has created at least short-term scope for a reshaping of previous desires and apparent certainties, and a rising up of new forms of enchantment from the ruptures created by the earthquakes.

3.5 Postsecularity, liminality, and event thinking

This chapter has traced the tectonic shifts, and resultant ruptures, which have led to the bubbling up of subjectivities of postsecularity. Postsecularity is not reducible to instrumental enlistment of religious capital into pseudo-governmental roles, neither is it a pragmatic tolerance of, or renewed openness towards, the capacity of theo-ethics to contest the moral terrain of neoliberal governmentality. Rather, we suggest the emergence of postsecularity as emblematic of deeper fissures in late capitalist subjectivity pertaining to the 'hollowing out' of existential life and the deformation of ethical desire. While neoliberal technologies of desire are effective in working through affective and emotional cadre (fear, hope, sacrifice, optimism – Berlant 2011), we note the bubbling up of ethical consciousness and a self-reflexivity of the shallowness of neoliberal enchantment, its conscious cruelty in its relation of self-to-self and self-to-others, is becoming viscerally felt. Critics might question wherever postsecularity could be merely a salve that placates these energies for more radical action – absorbing energies in feel good charitability. While such an account overlooks the capacity of ethical action to catalyse wider political mobilisations, postsecularity provides an analytic that is attuned to the nascent emergences and hybrid formulations of changing religious and non-religious belief, identity, and practice. For this reason, we seek to avoid couching analysis of postsecularity in blanket categories of 'progressive' or 'reactionary'; recognising instead the messy politics of enchantment and ethical formation at work in unexpected and familiar spaces of activism.

We wish to conclude this chapter by focussing on the different trajectories in and out of spaces of postsecularity. Much of the previous focus of research has focussed on experience-based postsecularity. Here the focus on phenomenology has emphasised the importance of dwelling in spaces of encounter and negotiation that produce and sustain subjectivities of postsecularity. This kind of analysis directs our gaze to the affective and emotional geographies that help to form ethical and political subjectivity. Through this conceptualisation, we can examine the liminal and transitional nature of spaces: ranging from the oscillations between the rhythms and habitus of neoliberal subjectivity and the affective, or spiritual, life of postsecular activism; to the experiences of discomfort in stepping in and out of thirdspaces of postsecularity alongside existing dominant territorialisation of belief, practice, and identity. Experience-based analysis of postsecularity is premised on situational and longitudinal engagement across lines of social difference and how this can be transformative in generating radically receptive and generous self-other identifications. Such an analysis is deeply contextual and based on performativity in situ and biographical analysis of positionalities and ethics religious and non-religious actors bring to spaces of rapprochement.

However, as illustrated in the example of Christchurch above, an event-based analysis of postsecularity helps us think differently about the formation of ethical subjectivities and associated tenets of receptive generosity, rapprochement, and hopeful enchantment. Event-centred thinking focusses our attention on the formation of subjectivities of postsecularity – in other words, the ways in which people hold fidelity to a particular event that shapes their ethical, political, and theological commitments. Events encompass the seemly mundane, for example, the daily encounters across lines of difference in public spaces but can also address the collective scale of tragedy and loss. Recent geographic theorisation of event has been applied to disaster spaces to reveal the affective and embodied impact of subjectivities. Drawing on Badiou's theory of the event, Cloke *et al.* (2017) highlight ways in which the Christchurch earthquakes of 2010 and 2011 ruptured the established order of things by producing a 'distinctive space in fidelity to the event [which] has the potential to unleash new beginnings and imaginations' (p. 69). Focussing on events shifts our focus onto the ways subjectivities are 'caught in a process of becoming' in and through events (Dewsbury 2000, 40): as Dewsbury further argues '[e]vents unfold in the relations of bodies, both organic and inorganic, human and non-human, and physical and ideal, where it is the relation that is important' (Dewsbury 2003 in Simpson 2017, 4).

As Cloke *et al.* (2017) suggest, use of the concept of event permits us to concentrate on how the everyday mundane discourses and practices of affected parties are performed in ways that are faithful to the event. While Badiou's original concern with the event was to trace the emergence of truth processes, and subjects' commitment to these emerging truths (see Fisher 2005), the notion of fidelity also stretches to more improvisational possibilities opened up by events, recognising that emergent and unanticipated lines of flight also contribute to processes of post-event resistance, and open up new pathways to change. Cloke *et al.* conclude:

> The event, therefore, is not simply a mappable moment, but an occurrence that exists and pervades as individuals show fidelity to it. Thus, in convoking and proclaiming truths in the name of the event, particular improvisations and ways-of-doing are both brought to the fore and made possible in a reordering that was previously neither thinkable nor graspable.
>
> (Cloke *et al.* 2017, 78)

Thinking through events in relation to postsecularity, then, is more than revealing key moments in religious and non-religious biographies, although this is clearly important; rather, this focus of analysis is on how subjects remain faithful to events, how events are generative of creative possibilities through affective mobilisation, and how these creative collectivities can rupture and rework existing government and non-governmental spaces of ethical action. This perspective helps reveal how subjectivities and practices are produced and sustained through events, and (re)configure the conditions of possibility – something that can be seen in relation to post-Occupy subjectivities (see Chapter 5). Understanding protest spaces as events therefore helps consider the ways in which event-spaces make possible collaborations that have been obscured through longstanding religious-secular distinctions.

A similar dynamic can be identified in the mobilisation of subjectivities of postsecularity against the white supremacist marches in Charlottesville, Virginia, on the 11–12 August 2017. Following a night of intimidation by torch-bearing white supremacists who encircled a small group of counter-protesters by the statue of Thomas Jefferson and trapped more than 600 counter-protesters inside St. Paul's Memorial Church, Jenkins (2017) documents the prayerful resistance of more than 80 clergy who in the early morning went to the gathering area of the 'Unite the Right' rally:

> As clergy lined up to face the field, a row of camp-clad militia members, draped in long guns, stared back at them. After a while, the clergy kneeled and prayed, one by one. Then they sang together. Some began to march

between the faith leaders and the militia, calling them 'weak' and shouting 'you really believe that?' in their faces as they sang. 'We got a lot of vitriolic slurs,' Wispelwey said. 'Most of them homophobic.' As Black Lives Matter protesters amassed nearby, a gaggle of white nationalists congregated behind the line of armed men to sing their own white nationalist songs. In response, the clergy began chanting 'love has already won.' The Black Lives Matter protesters quickly joined in. 'We sang, "Love has, love has, love has already won" ... even in the face of those [guns],' Wispelway said.

'My dominant impression was the incredible courage of the Black Lives Matter people who set up right near where the clergy set up – so in the very center of the gathering crowd,' McLaren said. 'I watched bottles and sticks flying, and then canisters of smoke. And I don't know what was pepper spray, what was mace, what was smoke bombs ... I saw Black Lives Matter signs being taken from protestors. They would pick those signs up and stand their ground.'

(Jenkins 2017, N.P.)

With little to no police protection, the event encapsulated the 'risky' and deviant hopefulness of religious and secular activists who put their bodies on the line to challenge white supremacy. The symbolism of clerical dress, music, and prayer – as well as physical blocking of entrances – sought to interrupt the logic of racist violence, and the violent clashes that ensued led to improvised affinities and partnerships that affirmed new, and existing, solidarities between clergy, Black Lives Matter representatives, and members of the antifascist movement (Democracy Now! 2017). Members of Antifa who were initially suspicious that faith leaders where there to 'protect the white supremacists' (Jenkins 2017), intervened to protect 20 clergy who had blocked an entrance to the 'Unite the Right' rally: 'The anti-fascists, and then, crucial, the anarchists, because they saved our lives, actually. We would have been completely crushed, and I'll never forget that' (Cornell West in Democracy Now! 2017).

This example illustrates the janus-faced nature of postsecular collaborations in the public realm in the USA. Christianity – in the form of a nostalgic ideology (Whitehead *et al.* 2018), an iconographic repertoire (for example, the use of crusade crosses; Tharoor 2016; Washington Post 2017), and prominent leadership (for example the role played by evangelical Jerry Falwell Jr and others; Green 2017) – became embroiled and enrolled into an event that defended white privilege, provided support for the Trump administration, and selectively remained silent in its stance towards white violence. While fault-lines between white nationalism and Christian conservativism remain intact (see Berry 2017), Charlottesville highlights the ambivalent ways in which some quarters of white Christianity might be wooed by the far-right into a resonance machine that marshals *ressentiment* towards the queer, immigrant, and multicultural 'other' who are charged with the loss of Christian heritage (see Bialecki 2017). Yet, the events in Charlottesville also show a more radical strand of Christianity within the US – including the longstanding linkages to the civil rights activism, but also more expressions of monastic and neo-anarchic faith (Claiborne 2006; Christoy-annopoulos 2011), that is a prominent supporter of and participant in anti-racist protests through the use of religious buildings, actors, and vigils.

Events of loss and tragedy can instigate powerful moments of outpouring of ethical values. On 14 June 2017, 72 people who lost their lives in the Grenfell Tower fire, a council owned high-rise housing block in the London Borough of Kensington and Chelsea. Much has been written about Grenfell, including the ignorance towards health and safety concerns of residents (Grenfell Action Group 2016), the moral bankruptcy of outsourcing responsibility across the private and public sector that has become all too common in neoliberal govern-ance, as well as the public and political representation of poverty, inequality, and economic insecurity in the aftermath (Shildrick 2018). In *After the Fire* (written by Alan Everett, Vicar of St Clement and St James), and *After Grenfell: The Faith Groups' Response* (a Theos report based on 30 interviews with local faith representatives – see Plender 2018) it becomes clear that churches, mosques, synagogues, and gurdwaras were at the heart of the immediate response to need, by providing practical, emotional, and spiritual support. These accounts along-side others (Sherwood 2017) detail: the interfaith collaboration organised through the space of the church and the local Mosque; the immediate suspension of moral and religious differences that might divide; and the capacity of public lament and bereavement to unite diverse religious and secular constituencies – bringing about new understandings of, and even temporary suspension of tradi-tional religious-secular distinctions. At least 15 centres run by faith communities provided aid including:

> [e]vacuation areas, receiving, sorting and distributing donations, offering accommodation, drawing up lists of the missing, supporting emergency ser-vices, patrolling the cordon, providing counselling and supporting survivors seeking housing. In the first three days alone at least 6,000 people were fed by a range of faith communities.
>
> (Plender 2018)

Bartolini *et al.* (2017) note that the 'spontaneous expressions of compassion were often couched in spiritual terms, including the creation of shrines, the use of candles and the invocation of God and angels' which not only enfolded cat-egories of the sacred and secular space, but also:

> undermine[d] the separation of different kinds of religious spaces from one another. More than this, it suggests that spirituality can lie beyond the formal spaces and practices of religion.
>
> (Bartolini 2017, 8)

The event also changed acceptability of religious voices and actors in the public. Sherwood (2017) tells the story of a local Methodist minister, Mike Long, who until the fire rarely wore a dog collar. However, at 4.30 on the morning of the fire he put it on and has never taken it off at any public engagement since. 'Now', he says, 'my role is much more public, and I need to be identifiable' – suggesting a new valuing and relaxation regarding public religious identity. It is unclear wherever we will eventually see the same level of acceptance of other public religious symbols, such as the hijab, emerge as part of the wider, national healing and cohesion that could come out of the awful trauma of this event. In his account of the healing and cohesion facilitated by faith groups in the aftermath, Baker (2017b) notes that the public performance by churches, mosques, and other secular institutions and individuals played a key role in denouncing the 'blind logic of managerialism' and its cost-cutting algorithms that rationalised the lack of sprinklers and the use of combustible cladding. The groundswell of ethical response, both nationally and within the local community, worked in part to restore webs of connectivity and hope, as the event instigated new imaginations of social value that extended across and through time and place. The aftermath of the Grenfell Tower disaster shows postsecularity across religious and secular differences arising spontaneously in the affective response of compassion, grief, and solidarity; as well as the suspension and reformulation of prevailing sacred/secular distinctions. What has been revealed by Grenfell has come to manifest a '*truth process*' (Badiou cited in Cloke *et al.* 2017, 72), producing community subjectivities marked by politicised hope and collective struggle against local authorities, and a national clarion call of disenchantment that refutes the ethical credibility of neoliberal technologies of outsourcing and 'efficiency savings'.

Note

1 We would like to acknowledge our debt to Justin Beaumont for initial conversations on the spiritual negation and ennui of neoliberalism.

4 Spaces

4.1 Introduction

This chapter is the first of three chapters examining the variegated geographies of postsecularity in different global, social, and political arenas. We begin by mapping the different spatial contexts in which postsecularity is emerging and the heterogeneous concerns being mobilised. Formations of postsecularity are highly contextual, and to illustrate this we compare the arenas of social welfare and environmental activism as constitutive fields that coalesce different sets of discourses, practices, and subjectivities. In each arena emergent postsecularity is shown to be generative of, and shaped by, different capacities for rapprochement and configurations of the religious and the secular. To develop this further, we examine the 'politics of curation' at work within emergent spaces of postsecularity through two focussed case-studies: faith-based drug and alcohol treatment and recovery, and emergency food provision in the UK. These empirical case-studies show how the three modalities of postsecularity – receptive generosity, rapprochement, and hopeful re-enchantment – are working out differently in different spaces. By foregrounding the configurations of religious and non-religious curation, we provide a more grounded assessment of the kinds of politics and ethics emerging in and through spaces of postsecularity.

4.2 Spaces of postsecularity

Recent years have witnessed a proliferation of scholarship exploring the notion of the postsecular in a variety of spatial contexts. The city as a unit of analysis has attracted most attention for postsecular scholars on three grounds. First, cities are sites characterised by diverse sets of ethnic, cultural, and religious plurality, and scholars have attempted to grasp the complex urban micro-politics of cosmopolitanism, conviviality, and (non)encounter, as well as how immigration and postcolonial identities in Western liberal democracy might challenge established structures of secularity and ideologies of secularism (Gorski and Altınordu 2008). Second, social marginality is most clearly visible and intensely concentrated in sites of the urban, which is reflected in the socio-spatial imagination of religious and secular individuals responding to inequalities. Third, cities are usually where we see scale and organisation of faith-motivated actors, both within and beyond faith boundaries (Cloke and Beaumont 2013). However, recent scholarship has begun to highlight the emergence of postsecularity in a range of other spaces, including the rural (Jones and Heley 2016; Meyer and Miggelbrink 2017) and the spatialities of transnational and global justice movements (Mavelli and Wilson 2016). Focussing on the variegated *geographies* of postsecularity, our attention is on the particular sites, spaces, and practices where diverse religious, humanist, and secularist voices come together in a dialogic manner and enter into a learning and experimental process in which secular and religious mentalities can be reflexively transformed (see Cloke and Beaumont 2013; Herman *et al.* 2012; Cloke *et al.* 2013, especially pp. 22–23). Some of the key spaces where emergent postsecularity has been recognised, and which have already been alluded to in Chapter 1, are as follows:

Spaces of care, whether they exist for homeless people, asylum seekers, victims of trafficking, victims of indebtedness, and other socially excluded groups, constitute a key arena of discourse and praxis for emergent postsecularity (Cloke *et al.* 2010; Cloke *et al.* 2013; Williams 2014). These are spaces where the occurrence of obvious social need is met by an affective response from the collective ethical and political conscience. Regressive political-economic changes are creating deepening injustices, shrinkage of welfare and rising levels of impoverishment, indebtedness, and exclusion. These needs can prompt new interventions and crossovers between faith-motivated and other actors, creating spaces of intensity, energy and hope for just alternatives. Faith-based organisations have been specifically identified as serving as practical devices through which diverse religious and secular motivations accrete around shared ethical impulse and crossover narratives (Cloke 2011a).

Spaces of direct resistance and subversion represent sites where rapprochement across religious/secular divides focusses on coming together in the deprived, disempowered, and most marginal spaces of the city to politically organise or join in with direct provision for the socially excluded. This involves working inside government funding schemes, but in such a way as to undermine or deflect the neoliberal politics involved, as well as establishing alternative services using charitable and voluntary resources in ways that contravene the hegemonic ideology. Rapprochement in this context is explicitly motivated and performed in the light of a critique of unjust socio-economic and political policies of neo-liberalism, combined with a desire to pursue philosophies and objectives of care which contravene the state's insistence on responsible neoliberal subject-citizenship (Cloke *et al.* 2010). This is clearly evident in spaces emerging to meet the needs of people for whom the state has chosen to withdraw its support (for example, single homeless people or people navigating the draconian asylum and immigration system). These spaces are often characterised by a moral contestation over dominant welfare constructions and programmes – with regard to restrictive eligibility, conditionality, un/deservedness, just rewards, and dependency. In the UK, several groups have been established, or have intensified their activity, in response to the severe welfare and housing benefit cuts introduced since 2010 as part of austerity programme of Coalition, and then Conservative government. Williams *et al.* (2014) highlight the example of Zacchaeus 2000 (Z2K), a London-based anti-poverty charity that helps low-income households affected by welfare reform and debt, including those forced to relocate due to cuts and caps on housing benefit. Z2K was founded in the early 1990s by Christians who refused to pay the Poll Tax and worked with other poll tax defaulters entangled in the court system. Z2K attempts to mitigate government policy and expose the 'unfairness in the law, legal and benefits system' (p. 2809). By appropriating legal technologies usually out of reach of vulnerable groups, Z2K's work embodies a political and politicised space of contention that reclaims notions of 'fairness' in judicial proceedings. Z2K combines localised support for individuals with advocacy, lobbying, and protest, and has been active recently working with the UNITE Community trade union to organise public demonstrations and protests in London against austerity-led welfare reform.

Spaces of ethical identity involve campaigns around particular ethical tropes at city level (see, for example, Clarke *et al.* 2007 on Fairtrade; Darling 2010 on sanctuary movement; Jamoul and Wills 2008 on living wage) which reflect a range of religious and other interests brought together to express identities and values, within which lie significant points of ethical convergence between theological, ideological, and humanitarian concern. Howson (2011), for example, discusses the 'We are Bradford' campaign that emerged as a public attempt to contest a demonstration by the fascist group, English Defence League (EDL), in Bradford 2010. It is an example of localised collaborations across Christian groups and Unite Against Fascism (UAF), that comprised union activists, local Muslims, students, and members of socialist and anti-fascist groups. Postsecularity can also been seen in the sanctuary movement in the USA, which has developed unlikely alliances and rapprochement between religious motivated actors and non-religious justice activists (Jones 2008). Through a rich ethnographic analysis of the New Sanctuary Movement in the USA, a network that seeks legalisation and rights for undocumented people, Yukich (2013) provides an account of the challenges and competing interests involved in alliance building, or what we might call postsecular rapprochement. She highlights the strategic subversion of right-leaning religious discourses of 'family values', and how the New Sanctuary Movement utilises the religious discourse of 'One Family under God'. However, Yukich (2013) details the ambivalences of this space: namely, the partiality in focussing more on mixed-status families than undocumented individuals, as well as the reinforcement of the precarious status of undocumented individuals that discloses the gendered and racialised constructions underpinning notions of hospitality (see see Ehrkamp and Nagel 2014). She also highlights the tensions experienced in seeking to initiate alliances with religious congregations. Significant difficulties in recruitment within evangelical Protestant Christianity resulted in the interfaith character of New Sanctuary Movement mostly encompassing 'mainline Protestants, liberal Catholics, and Reform Jews' (Yukich 2013, 198).

Spaces of protest occur where pluralistic sensibilities and horizontalist organisation of recent social movements – such as Occupy Wall Street, Taksim

Gezi Park, and the Arab Spring (see Cloke *et al.* 2015; Dabashi 2012; Mavelli 2012; Barbato 2012) – are marked by an explicit 'crossing over' of religious and secular narratives, symbolism, practices, and performances in public space. Slessarev-Jamir (2011), for instance, offers a series of case studies of Christian, Jewish, and Buddhist justice movements active in: congregational community organising; worker justice; immigrant rights work; peace-making and reconciliation; and global anti-poverty and debt relief. While we develop these themes in more depth in Chapter 5, here we offer one example of how elements of postsecularity are emergent in recent protest spaces. The 'Moral Monday' movement of civil disobedience in North Carolina, began on 29 April 2013, when 17 church ministers and leaders of the National Association for the Advancement of Colored People (NAACP) were arrested inside the North Carolina General Assembly building in Raleigh for protesting against the punitive welfare, education, health cuts, and racially discriminating voting-ID laws passed by the Republican administration who control the general assembly. In the following two months, the NAACP, organised through the churches, mobilised a state-wide campaign of nonviolent civil disobedience where over 700 people were arrested for following suit. The protest was primarily clergy-led yet attracted support and active participation from: Trade Unions/Workers Assemblies; War Veterans; schools; Democracy activists; Jews for Justice; and various Christian denominations – even some evangelicals. The result was to gain a pragmatic acceptance from some quarters which would be otherwise at odds with the moral identification with religion. Since 2013, the Moral Monday movement, can be considered to constitute a 'counter-resonance' machine, bringing together divergent ideologies and religious, ethnic, sexual identities (Phelps 2013), to disrupt the direct support, or acquiescence, among conservative evangelicals in rural North Carolina for the Republican administration. Phelps writes:

> The various religious and non-religious dispositions and issues represented cannot be lumped together under a common, ideological banner, one that would attempt to smooth out the tensions that exist among individuals, groups, and causes. Rather, these resonate on the basis of a shared 'spirituality', a mutual ethos that allows divergence and convergence in a specific space.
>
> (2013, para 7)

More recently, the advent of the Trump administration with its Muslim ban and increasingly harsh immigration policy, combined with North Carolina's government anti-LGBT legislation and the high-profile police killing of unarmed black citizens, has further served to unite disparate interests and identities to the extent that in 2016 North Carolina seemed to be the epicentre of mass protest in the USA (Blest 2017).

Spaces of local reconciliation and tolerance come into being where individuals and groups work across, or least problematise previous divides involving inter-religious, anti-religious, or anti-secular tensions. Whilst these spaces can include government-enlisted attempts, more grassroots and self-organised forms seem more effective sources of reconciliation. These are everyday sites of embodied performance of identity – religious or otherwise – in which local lived spaces come to represent the potential for new formations of tolerance and assimilation in place of sectarian divides (see Garrigan 2010). These spaces are characterised by deliberate modes of organising that emphasise interfaith dialogue or co-presence. In a study of religious peacebuilding in Ambon Island, Eastern Indonesia, Al Qurtuby (2013) details the role local Christian and Muslim leaders played in fostering spaces of interfaith dialogue and reconciliation after the initial outbreak of communal violence in 1999. In the absence of large-scale peacebuilding initiatives, Al Qurtuby examines the significance of two interfaith peacebuilding groups: *Provokator Perdamaian* – a diverse network of Christians and Muslims from various backgrounds and professions – and *Tim 20 Wayame* – highly localised collaboration between villagers and religious leaders in the conflict zone of Ambon. Both initiatives recognised the need for collaboration across religious boundaries to disrupt theologies of violence and developed dialogical spaces to reflect and practice common values such as justice and compassion inspired by faith traditions. Other examples include: interfaith education (McCowan 2017; see also Hemming 2011; Watson 2013); reconciliation and forgiveness through integrated schooling in Northern Ireland (McGlynn *et al.* 2004); community growing initiatives (Kimberly *et al.* 2004); economic cooperatives in Israel-Palestine (Voinea 2014; Nusseibeh 2011; Maoz 2004; Arab American Tribe 2015); Ecumenical community projects in Glasgow that enable Protestants and Roman Catholics to work together to meet local needs and provide opportunities for people to discover 'common goods' (Harvey 2011); and finally, economic cooperatives such as Delicious Peace, Uganda where religious – predominantly Jewish, Muslim, and Christian – and nonreligious farmers self-organise, share resources, and profits into a variety of public health and education projects (see Auerbach 2012; Lidman 2014)

Spaces of voluntaristic or charitable cross-subsidy and solidarity include ethical projects aimed at welfare, care, and social justice that depend for financial backing and voluntary labour on people from other parts of the city, usually the more affluent suburbs where social need rarely presents itself. By providing flows of resource into spaces of care in the marginal areas of cities, these sources of suburban social and spiritual capital develop socio-spatial connectivities that unsettle not only perceived religious/secular boundaries, but also territorial ones. One example of this is re:Source Bristol, a cross-church partnership of people relocating to margins of the city and in so doing unlearning cultural and religious habitus of presumed (often prejudicial) certainties of religious belief and ways of working (Cloke and Pears 2017a, b).

4.3 A spatial analytic of postsecularity?

What these variegated spaces of postsecularity do, shaped by different configurations of discourse and practice, is to present opportunities for highly contingent formations of receptive generosity, rapprochement, and enchantment. We understand this to require a new spatial analytic. Put plainly, each arena is generative of, and shaped by, specific ethical and political subjectivities. Space here is not simply a container for action, a passive backdrop on which discourses and practices coalesce. Rather, we suggest space is active as a constitutive force in the emergence of postsecularity. This involves recognising that the manifestation of religion and secularity will vary between different places, and on different scales. But more than this, it entails a recognition that space plays an integral part in the nascent relationalities of receptive generosity, rapprochement, and enchantment. The ethical capacity to respond to injustice, and the desire to enter into rapprochement with others, is mediated by spatial relations of proximity and distance (Lawson 2007; Massey 2008; Popke 2007) as well as an individual's physical, social, and emotional constellation within an established habitus of theological expression. Building on recent geographic scholarship on ethics, care, and responsibility (Young 2004; Lawson 2007; Massey 2004, 2008; also see Noxolo *et al.* 2012), analysis of postsecularity focusses on the critical role of mediation in the opening up – and closing down – of ethical and political subjectivities; recognising that connectivity and propinquity are not guaranteed to produce ethical sensibilities, and that technological devices offer routes to transcend the extension of care across distance. Postsecularity helps us analyse the inherently spatialised relations between care, ethics, and politics, particularly in cities where the phenomenology of need is usually most visibly pronounced. Here we acknowledge the visceral desire to 'do something' (Cloke and Beaumont 2013) is an affective force that emerges as an embodied response to the modern life of cities – which have the simultaneous capacity to produce indifference and a blasé attitude to suffering (Simmel 2002 [1903]), as well as everyday senses of togetherness, encounter, and conviviality (Wilson H 2014). Postsecularity does not emerge on a blank canvas and it is essential both to examine the highly contingent nature of historical situatedness, and to recognise place-based formations of postsecularity that emerge through longstanding resonances or affective atmospheres that are tied to the histories of a place (Ivinson and Renold 2013) and shape the localised proclivities for ethical and political collaboration. Ultimately, our conceptualisation recognises the spatial *processes* at work in the formation of emergent subjectivities and spaces of postsecularity. We build on established approaches to feminist ethics of care to conceptualise ethical subjectivity as produced through *relationalities* with human and nonhuman others – thereby challenging liberal visions of a bounded, autonomous moral subject (Braidotti 2006; Lawson 2007; Barnett *et al.* 2005). Our conception of postsecularity as a site of being-in-common foregrounds the movement from extant comfort zones to the space of rapprochement – a thirdspace where previously held beliefs and ways-of-being are negotiated. This helps us reveal the reterritorialisation of people, ideas, flows both between and within institutional and non-institutional theological contexts, and movements within and beyond forms of secularity.

For this reason, we do not assume these case-studies to be universally revealing but rather suggest they give an indication of the formation of postsecularity in specific contexts. We are interested in the dialectical relations of spaces and subjects that are co-constitutive of particular kinds of networks, assemblages, and affinities. To illustrate this point, we now briefly examine the arenas of social welfare and environmental activism as constitutive fields that coalesce different sets of discourses, practices, and subjectivities. Our purpose of this framing is to explore the 'politics of curation' within spaces of postsecularity, which helps foreground analysis of the different assemblages of discourse, disposition, and practice that sustain collaborations. Specifically, the idea of curation refers to the assembling of a space in such a way as to allow diverse influences and agencies to come together without excessive control, ensuring that each can make its own contribution to the whole.[1]

Neoliberalised spaces of welfare

This is an arena where the mutual translation of religious and secular ethics has been made most clearly visible through existing academic scholarship on postsecularity. Much has been written about how the iconic signification of injustice – homelessness, asylum, trafficking, food poverty – has led to a groundswell of desire to 'do something' among religious and secular voices resulting in a diverse set of religious, ideological, and humanist positionalities coalescing in the pursuit of common ethical impulses (Cloke and Beaumont 2013; Baker 2012). In the broad area of welfare and care, emergent postsecularity has been opened out through three manoeuvres. The first concerns the renewed openness among local government to collaborate with faith-motivated groups, initially out of a recognition of the unique resources associated to religious and spiritual capital (cohesion, representation, service provision – see Baker 2012; Dinham and Lowndes 2008), and more recently perhaps, as part of an austere-shaped pragmatism to meet the needs of citizens without sufficient resources. These spaces are initiated and curated predominantly by secular actors but open out spaces of mutual engagement and collaboration across religious and secular difference. The second manoeuvre concerns the changing theologies and ways-of-working among religious organisations and actors in welcoming secular and non-religious voices to work collaboratively towards social justice projects. This is most clearly seen in the arena of homelessness (Cloke *et al.* 2010) and food-banks (Williams *et al.* 2014). In these spaces, emergent postsecularity emerges in and through spaces directly or indirectly tied to the work of faith-based organisations. The arena of welfare and care presents obvious opportunities for faith groups to infuse theo-ethics in local settings, eschewing 'conversion-oriented' motivation and defensive postures that 'badge' activities as exclusively religious in fear of a dilution of 'faith element'. Instead, there has been a bubbling up of traces of a different kind of religious and secular mix in welfare settings, characterised by new kinds of reconciliatory and partnership ethics based on an overt receptivity to religious and ideological difference.

The last manoeuvre refers to the emergence of broad-based community organisation which has not been produced through overtly 'religious' or 'secular' space, but rather constitutes a thirdspace that deliberately blurs religious and secular boundaries. The work on London Citizens (Jamoul and Wills 2008), but also see analysis of earlier faith-led broad-based community organising in Manchester at the height of New Labour's urban regeneration policy through organisations such as Community Pride (Baker 2009), is an instructive example of the ways a diversity of churches, Mosques, schools, trade unions, and community groups collaborate to address issues such as the Living Wage. The Citizens UK model offers a highly curated device that deliberately sets aside issues of moral and ideological difference (abortion, women's, and gay rights), to focus on shared justice concerns. Yet in the process of collaboration opportunities emerge for dialogical learning where faith-based and secular interests learn from each other and set aside prefigured and often prejudicial attitudes.

Environmental activism

The environmental arena illustrates the contested nature of postsecularity, namely, the difficult and asymmetrical relations experienced by religious and non-religious activists. Over recent years there has certainly been a rise in what might be regarded as postsecular environmentalism, embodied in the turn towards deep green spirituality and the loosening of religious-secular distinctions (Wickström and Illman 2015). However, the literatures in this field often do not fully appreciate the distinct challenges facing the mobilisation of religious groups into environmental action. For example, there remains a hesitation and resistance towards religion in some quarters of environmental activism, stemming from the historical legacy of religious involvement in environmental discourse – particularly Christian theologies of dominionism. This tradition haunts dominant discourse regarding religious environmental activism. In her ethnography of 2009 Climate Camp in London, Nita (2014) notes that camp rules were explicit that no religious groups should be allowed on site, which 'was explained by organisers as a strategy for maintaining unity' (p. 233). According to Nita, non-faith activists built on a common narrative that tended to exclude those who identified with traditional religious institutions (although she notes this took the form of identifying with groups that had experienced political or religious oppression, for example Diggers). She argues that Climate Camp and to a lesser extent the transition movement, have held an anti-religious ethos, which in turn has led to faith activists feeling obliged to act as 'closet Christians' (p. 234). Effort is thus made, it seems, to regulate secular space, circumscribing collaboration to those of broad adherence to a core identity.

While there are countervailing signs that environmentalism *is* reaching out to religious groups or at least tolerating religious voices for purposes of collaboration, there has been difficulty translating some secular and scientific discourses of climate change into religious congregations. Faith-motivated environmentalists are marginalised and on the fringe of institutional religion. Nita (2014) describes the liminal identification experienced by religiously – identifying environmentalists: namely, the difficult experience of 'coming out' as a Christian in the Green Movement as well as that of 'coming out' as a Green in their church' (p. 235). The environmental arena might therefore outline a space of postsecularity defined by *asymmetries of translation*. Desire to respond to climate change captures religious actors, but this is not explicitly valorised as such by lead protagonists within congregations. Perhaps as a result, faith-motivated environmental activists explore opportunities beyond the institutional religious frame in their locality, channelling involvement in a host of para-congregational organisations (Christian Ecology Link, GreenSpirit, Isaiah 58, Student Christian Movement, Operation Noah, A Rocha, Christian Aid, and others). Involvement and membership in such organisations have grown through wider online activism. For example, London Islamic Network for the Environment was established in 2004 by Muslim environmental activists, changing its name to Wisdom in Nature (WIN) in 2010. Although there are few active members present in Climate Camp, there is a much larger online following (Nita 2014).

Umbrella organisations seeking to mobilise religious communities in environmental activism have helped curate new spaces and alliances among religious and secular motivated actors. The example of Transition Towns is indicative of how people of religious faith play significant roles in shaping and leading the movement but are less explicit about their religious motivation for doing so. Members of Christian Ecology Link (CEL) were integrally involved in community building work of Transition Towns, and as an organisation CEL offered support to the Transition Towns Network in the form of a flanking campaign called Churches in Transition (launched in 2009 – see Nita 2014, 230). Yet those involved trying to drag religious institutions and actors into these spaces experience difficulties in overcoming ecological illiteracy and motivation among religious individuals and congregations. The perceived lack of salience, or emotional pay-off, of environmental activism is considered one of the barriers in motivating religious people in environmental activism. Compared to the explicit theo-ethical motifs to 'care for the widow, sick and orphan' and 'when I was naked, you clothed me, when I was hungry, you gave me something to eat'; it is easier to actualise theo-ethical concern for justice on particular (often localised) issues such as homelessness, food poverty, and debt advice.

Despite these limits, postsecular-styled spaces of rapprochement in the arena of environmental activism are conducive to a hopeful enchantment – soliciting new expressions and identifications that blur traditional divides between the religious and the secular. Nita's examination of Christian and Muslim identities involved in Climate Movement and Transition Towns movement in Britain is again instructive here. Participants expressed faith primarily through ecological concerns and vice versa – ecological concerns framed and motivated by faith: 'Their faith had become imbued with ecological beliefs and practices, or perhaps hybridised with dark green beliefs and practices through their involvement in the Climate Movement' (2014, 229). Spaces such as Transition Towns and Climate Camp, then, might be re-read as liminal spaces of rapprochement where participants identifying as Christian, Muslim, Buddhist, Druids, and Pagans as well as religiously unaffiliated activists, translate religious and spiritual motivations.

Building on Bron Taylor's (2010) 'dark green religion', the claim here is spaces of rapprochement in environmental activism have a greater tendency for activists to merge their green and faith identities. Furthermore, as Taylor (2010) and others (for example, Bennett 2001) have noted, such collaborations might offer a re-enchantment with the world characterised by biocentric, and not anthropocentric, concerns. As recent work (Smith *et al.* 2017; Blaser 2014) has highlighted, greater recognition of spiritual ontologies is becoming more acceptable for the conceptual palette of the secular academy. Theoretical turns towards ontological politics in environmental justice and development literatures have reframed recognition of indigenous perspectives in environmental activism – bringing a renewed openness to polytheist and nontheistic ontologies at work in ethical sensibilities towards nonhuman others. This has been accompanied with a spiritual turn in Human Geography scholarship more widely (Bartolini *et al.* 2017; Holloway 2013). As Dwyer (2016) notes much of geography scholarship has been driven by a 'modernist academic gaze, which ignored or suppressed the agency and salience of the sacred, in favour of approaches which include the religious and the spiritual in frameworks of analysis and explanation' (Dwyer 2016, 758). The influence of Durkheim-inspired situational approaches in geographies of religion, while bringing important insights on religious forms of identity and social formation, has meant 'theological ontologies and sensibilities of the religious and spiritual have been too often sidelined or ignored' (Holloway 2011a, 31). Emphasis has shifted onto how to give analytical space 'to those extraordinary forces that the faithful will *always* say move them to action' (Holloway 2013, 204). More widely, these moves can be plotted in similar trends in the growth of indigenous activism, the spiritual ecology movement and its renewed emphasis on spiritual values and motivations that underpin and are formative of care-full – and uncaring – relationships to the non-human.

From this analysis we can see that emergent postsecularity takes distinct forms in different social and political arenas. The arenas of social welfare and environmental activism present different constitutive fields of assemblage that coalesce different sets of discourses, practices, and subjectivities. The neoliberalised context of welfare and care presents multiple opportunities for rapprochement – where religious actors create space for non-religious voices to join forces in social justice projects, and the secular opens out space for faith-motivation most clearly seen in the incorporation of religion for 'ready-made' community representation, cohesion, voluntary resource, and service provision. In contrast, the environmental arena of activism is characterised by the difficulty often experienced by religious activists in collaborating in secular environmental groups or trying to incorporate ideas from the secular scientific world back into the congregational context to find unreceptive and resistant attitudes. There are various explanations for this that include longstanding anthropocentric imaginations of justice among many religious groups, specifically scriptural texts that seem to draw emphasis to commands to 'care for the poor' 'clothe the naked, comfort the widow' than environmental justice. Within this, there is often a synergy between what is problematically considered a lack of 'salience' within environmental action, and perhaps the costly demand environmental justice involves in terms of calling time on personal and institutional religious complicity in structural violence that characterises the fossil fuel industry. Within the environmental field, theological doctrines of dominion and stewardship, as well as historic lack of care given to the non-human environment by dominant forms of monotheist religious traditions, has led to barriers for religious groups to overcome in more secular-led environmentalist groups. However, as noted above, this might change through the more pluralistic and receptive attitudes towards indigenous ontologies in environmental activism, which might connect with 'dark green faith' to articulate more explicitly the nascent sacralisation of the non-human environment. We return to this theme in Chapter 7 when we discuss postsecularity in relation to a new politics of hope.

4.4 Grounding emergent postsecularity as a politics of curation: two case-studies

In this section, we provide two in-depth case-studies drawn from our empirical research in the arena of welfare and care.[2] They help illustrate the 'politics of curation' at work in different spaces of postsecularity – how the three modalities of postsecularity are constructed and contested 'on the ground'. We pay attention to the different characteristics of discourse, practice, and performance that produce and sustain ethical stances towards receptive generosity, rapprochement, and re-configurations of desire.

Subversive politics of postsecularity in neoliberal governance: the case of The Salvation Army's Hope House

This example draws on empirical fieldwork[3] in a homeless centre and drug and alcohol service run by The Salvation Army (hereafter TSA) in the UK (Williams 2015). Through it we demonstrate how postsecularity can open out ethical and political spaces *in* and *against* neoliberal governmentalities, and how a particular form of rapprochement across religious, secular, and humanist voices might negotiate and translate ethico-political subjectivities. Hope House (pseudonym) is a TSA Lifehouse providing entry-level emergency accommodation, and offers specialist maintenance, medical-based detoxification, and abstinence-based rehabilitation facilities for people with alcohol and drug problems. Located in a large English city, the 93-bed centre is funded directly by government through their Supporting People programme (see May *et al*. 2006) and works collaboratively with the local authority Drug Strategy Team (DST) responsible for commissioning and overseeing drug services in the area.

Hope House provides a fascinating arena in which to contextualise the emergence of postsecularity. We can identify at least four flows that have come to shape the proclivities for religious and secular voices to enter into rapprochement and coalesce over shared ethical concerns. The first, in line with some of the intellectual shifts identified in Chapters 2 and 3, concerns a wider shift in the practical theology of TSA. Founded in East London in 1865 by William and Catherine Booth, TSA set out as an evangelical missionary movement based on a quasi-military structure that promoted temperance and tied social assistance with an 'urgency to convert people to Christian ways of living' (Cloke *et al*. 2005, 389). In the last 20 years, TSA, particularly its social services arm, has moved away from a 'serve-you-to-convert-you' attitude that previously had made homeless service provision conditional on religious participation (see Snow and Anderson 1993). Instead, TSA have moved towards a theological stance of unconditionality, seeking to offer services 'without strings' – detached from any obligation to engage with religious activities (see Cloke *et al*. 2007 and 2012).

The second flow relates to changing paradigms towards person-centred care and spirituality in professional social care (Furness and Gilligan 2010). This has encouraged key workers to eschew their own metaphysical views and to harness individuals 'recovery capital' – recognising for some people religious and spiritual belief can play different roles in sustaining motivation for recovery, as well as internalise feelings of shame that can undermine recovery pathways. The third, and related, factor concerns the mainstreaming of holism and alternative spirituality in healthcare and addiction treatment (Frisk 2011; Greenstreet 2006). This mainstreaming recognises that secular modes of treatment and recovery have been co-constituted through religious organisations, (for example, the historic dominance of 12 step groups), narratives (of redemption, sin, and salvation), and practices (of confession, prayer) and regularly favour the DIY spiritualities of those in recovery (Dossett 2013). Spaces of alcohol and drug recovery, ranging from AA meetings to Buddhist mindfulness groups, are liminal spaces where people construct and (re)assemble a highly personalised mix of religious, secular, and scientific understandings of 'addiction' and recovery. Recovery space within Hope House therefore is characterised by an 'overt metaphysical/religious pluralism' (Connolly 1999, 185) within the discourses and pragmatics of care, and of recovery experience, making possible the negotiation and 'crossing over' between religious and secular ethics, beliefs, and practices.

Last, incorporation into the financial and regulatory frameworks of joined-up neoliberal governance has changed the modus operandi of Hope House. Technologies of contractualism, audit, and best value have meant Hope House regularly has to bid competitively for contracts – a practice that results in a degree of self-regulation, and (at least nominal) adherence to the desired philosophy and practices of funding commissioners. To maintain their rolling contract with local commissioners, Hope House has carefully managed its organisational image, for example, by curtailing overt displays of unwanted proselytisation, both on an individual and organisational level. Greater professionalisation has resulted in changing staff profiles. For example, trained and accredited key workers and drug counsellors, and combined equal opportunities legislation, has served to undermine the practicality of Christian staff. Equally, non-religious and secular drug and alcohol workers have come to work – perhaps reluctantly at first –

because TSA won the service delivery contract. As a result, the organisational space and agency of Hope House is characterised by overt and legislated partnerships between religious-secular individuals and organisations, which nevertheless create a rich assortment of motivations, discourses of care, and ways of working into a traditionally religious environment.

As such, Hope House has attracted a diversity of religious and non-religious workers who share a commitment to work with socially marginalised individuals. Just taking those staff involved on the drug programme, we encounter Salvationists, conservative evangelicals, Pentecostals, liberals, people of New Age and Buddhist faiths, agnostics and atheists of all ages and backgrounds, who each bring their ideas of what constitutes good practice, healing, development, and transformation. In this setting, postsecularity is intimately linked to the desire to work around and in some cases subvert the neoliberal metrics of deservingness and responsibilisation.

Table 4.1 Characteristics of religious belief among staff members on the drug programme

Name (pseudonym)	Staff role/position	Religious belief and (self)identification
Stephen	Centre manager	Salvation Army Officer, Pentecostal
Pam	Manager of Drug Programme	Salvation Army Officer, Salvationist (Corps)
Hannah	Senior detox nurse	Salvation Army Officer, Charismatic Evangelical
Katie	Detox counsellor	New Age, grew up in Conservative Evangelical church
Mark	Senior counsellor, Preparation and rehab unit	Atheist, Buddhist mindfulness facilitator and offered meditation classes for residents
Esther	Trainee centre manager	Salvation Army Officer, Evangelical
Dave	Senior detox nurse	Atheist
Sara	Rehab counsellor	New age, pagan, Goth
Gareth	Doctor	Anglican
Sharon	Rehab counsellor	Agnostic, New Age/Buddhist, practiced alternative therapy (massage, acupuncture and meditation) in Hope House
Tasha	Rehab counsellor	Agnostic
Neil	Head of outreach	Atheist
Joy	Receptionist	Salvation Army Officer, Pentecostal
Emily	Administrator	Conservative Evangelical
Molly	Former doctor	Conservative Evangelical
Richard	Preparation counsellor	Agnostic, grew up in Black Pentecostal church
Phil	Resettlement worker	Agnostic, Alcoholics Anonymous advocate

As part of a government-funded programme, Hope House's previous approach of permitting direct access was realigned to a 'referral-only' policy, which restricted eligibility to clients who were already engaged with 'mainstream agencies' (social services, official City Council Outreach Teams, or recognised Third Sector agencies) and who could demonstrate a 'local connection' to the city (see May *et al.* 2006; May and Cloke 2014). Equally, funding requirements associated with Supporting People and Drug Strategy Team programmes have placed strict time limits on how long residents could stay in the Lifehouse (target 6 months) and on the drug programme (6 weeks preparation; 10–14 days detox; and maximum 16 weeks rehab). Although staff recognised beneficial elements of professionalisation, for instance, in ensuring standardised quality of care in the area of dual diagnosis and mental health (Stephen, centre manager 12/1/10), there was concern that governmental technologies of audit, eligibility, and shifting funding regimes institutionalised a procedural ethics of care (standardised care plans and short-term targeted interventions) that often failed to meet complex and individual needs. These policy regimes represent a mixture of treatment identities (Fraser and Valentine 2008) that discursively construct service-users through notions of 'stability' and 'chaotic user' based on neoliberal metrics of self-regulation and responsibilisation (Monaghan 2012). Part of professionalisation in Hope House, and other accredited emergency accommodation providers, has entailed adhering to a standardised licence agreement that sets out the responsibilities and requirements placed on residents (for instance, prohibitions of substances on site) and the level and programming of service provision. Technologies of contractual governance have become increasingly prevalent in welfare, health, and crime policy, particularly in relation to drug use (Seddon 2010). The politics of contractual technologies instil a neoliberal problematisation of drug users as rational calculating risk-takers and choice-makers, in which non-compliance or failure to perform the behavioural expectations of the 'responsible service user' potentially becomes grounds for exclusion and other illiberal measures. Such ordered environments found in drug treatment and 'rehabilitation' spaces have been understood through a lens of social control (Wilton and DeVerteuil 2006), normalising 'unruly' subjects into docile, obedient bodies (Bourgois 2000), or as 'technologies of the self' that instil neoliberal values of risk management, self-help, and self-responsibility (Fairbanks 2009).

However, by attending to the neglected emotional geographies of staff, volunteers, and residents who inhabit and co-produce these regulatory spaces, we reveal how emergent forms of postsecularity, premised on the 'crossing over' of religious and secular voices, opens out the possibility of ethical and political spaces that rework and challenge neoliberal calculations of deservingness and individualisation of risk.

Postsecularity as subversive ethics

There are three distinct aspects by which the rapprochement of religious and secular actors has instigated subversive ethics in and against neoliberal governmentalities. First, the organisational identity of the TSA was strategically deployed as a representational device in negotiations with commissioners and Drug Strategy Teams. Stephen, the centre manager, explained that the 'Christian ethos of Hope House' gave him the flexibility and ethical rationale to challenge the government's referral-only homeless policy. When the referral-only policy first came in, Stephen and other TSA centre managers refused to adhere to the practice of turning people away without referrals and continued to operate a direct access philosophy as an outworking of their organisational identity. The centre provided six beds to individuals deemed ineligible under government criteria, using its own money from headquarters and private donations 'to meet any need as it presents itself' (Stephen, centre manager 12/1/10). By continuing to provide direct access facilities over the weekends and in the evenings, when most mainstream services close their doors, Hope House has opened out an interstitial space where neoliberal calculations of need, and associated attempts to delegitimise the philosophy of direct access, are tempered by postsecular notions of caritas.

Second, Hope House introduced new regulatory technologies designed to circumvent the intended policies and processes of neoliberal contractualism. Rather than the 'one strike and you're out' penalty obliged in the licence agreement of emergency accommodation providers, the management of the centre sought to instil a 'culture of forgiveness' by showing leniency to clients and finding new ways of addressing problematic behaviour. At first, these technologies emanated from the ordinary ethics of staff – religious and secular – choosing to find ways around evictions; but these practices later came to be formalised into new technologies such as the Alcohol Assertiveness Scheme, providing intensive key-worker support for residents trapped in the 'revolving door' of alcohol-related eviction. Frontline workers were aware that the procedural ethical code preserved in tenancy contracts leads to the eviction of those who break the rules, thus exacerbating the exclusion of individuals with chronic alcohol-related problems who recurrently break licensing agreements. As Stephen explains:

> [Y]ou're freed to do that within the Salvation Army system, so I can show mercy and be flexible in a way that I wouldn't be in other places I've worked. And that has to be down to the faith element because that's what the governing instrument is at the start – the founding roots of the organisation, permeating things through.
>
> (Centre manager 12/1/10)

Third, staff and residents played a key performative role in generating more hopeful spaces of collaborative care and empowerment – creating social roles and relationships that partially unsettled the stigmatised identities of the 'drug user' and neoliberal subjectivities of risk individualisation and self-responsibility. Residents praised staff who went 'an extra mile' to properly get to know them, and the administration of Hope House centre who used its own money to provide trips out and recreational activities such as football and cinema trips. Communal spaces such as the canteen where staff and residents ate breakfast, lunch, and dinner together offered an event-space of conviviality rather than differentiated subject positions of 'chaotic drug user' and 'professional staff'. We are not claiming that TSA is somehow unique in its service provision, nor assuming practices of sociality are characterised by symmetrical encounters devoid of power dynamics. Rather, we suggest practices of care and sociality serve as liminal spaces where the stigma or moral distance that can accompany professionalised care regimes is challenged, and new relations to alterity can be fostered. Turner's concept of *communitas* (1969) is helpful here to focus attention on the political significance of seemingly mundane spaces of sociality that overturn, albeit partially, established hierarchies between resident and staff. Temporary suspensions of cultural and social norms produced in these fleeting gatherings and liminal spaces of encounter seemed to be an unspoken but widely upheld practice at work in Hope House – shaping the proclivities for religious and secular actors to receptively engage with others and in some cases transition beyond judgemental attitudes.

Taken together, the examples above describe an intuitive resistance against the isomorphic pressures of austerity-driven bureaucracy that are galvanised in and through the crossing-over and mutual translation of religious and secular ethics and practices. We unpack this further by illustrating the ways in which postsecularity in Hope House was generative of, and shaped by, modalities of receptive generosity, rapprochement, and the hopeful reconfiguration of belief.

Receptive generosity

Receptive generosity in Hope House was manifest across three key registers: discursive devices and practices that enable the crossover of religious and secular motivation; recognition of a common ethical motivation; and through affective atmospheres that spur recognition of common ethical desire.

First, receptive generosity was evident in the ethical stances of atheist and Christian staff taking on criticisms of their own position and working from a reflexive position that is produced because of, and not despite of, this fusion of secular and predominantly Christian imaginaries and practices. This was partly enabled through the construction of crossover narratives and devices capable of holding together the combined discourses and praxis of secular and religious actors. As noted above, landscapes of addiction treatment have witnessed the mainstreaming of spirituality in healthcare, as part of a wider shift in the sociology of medicine from a paternalistic benevolent ethos in the delivery of professional care to an increased autonomy for, and by, those using social services (Greenstreet 2006, 24). Person-centred care has been accompanied by a renewed openness to non-Western spiritual practices as part of secular treatment programmes, which could also be seen in the pragmatics of care in Hope House, with several staff specialising in aromatherapy, acupuncture, massage, and Buddhist philosophy of mindfulness. These activities were made freely available and seemed popular among residents. The fact that these practices co-existed alongside optional Bible study classes and prayer groups, and fellowship meetings of Alcoholics Anonymous and other Higher Power groups, suggest an increasingly hybrid therapeutic space operates within Hope House, where facilitators and service-users appropriate and experiment with blurring the boundaries of scientific, religious, and therapeutic discourse and praxis (see Frisk 2011).

If the pragmatics of care in Hope House can be characterised by an overt metaphysical plurality, it is also important to note how staff and residents interacted with these often-blurred encounters. What was most noticeable in therapy groups was an attempt by staff – religious and secular – to put aside their own secular or religious perspectives so as to respond in ways that enhance the resident's capacity to engage receptively and generously with the world. This is a clear indication of the *receptive generosity* across lines of religious and secular difference. For instance, atheist staff sympathetically engaged with religious and alternative spiritualities – whatever the worldview of the client – in order to harness motivation for recovery in a way that respected alterity (Coles 1997). Equally, religious staff came to utilise 'secular' understandings of addiction and bracket out aspects of their own beliefs in order to work effectively with clients. These blurring encounters between religious and secular narratives of recovery came to be appropriated by the individual agency of residents.

Residents expressed a variety of different understandings and practices relating to their engagement in the treatment programme. Indeed, the individual agency of residents came to co-constitute, challenge, or otherwise reshape the postsecular conditions. Take, as an example, participation with Buddhist practices of mindfulness and the Twelve Steps of Alcoholics Anonymous fellowships. Some residents openly accepted these practices and engaged to varying degrees with their philosophical and religious traditions and meanings. One of the detox residents noted the need to 'find your own god … whatever you want to call it' and how 'you can develop that for yourself' (Colin, interview 20/8/10). Others saw the therapeutic value of these practices but detached them from their metaphysical signification, preferring instead to engage with a more individualised notion of Higher Powers and mediation – a move some have argued to reflect a post-modern negotiated spirituality (Dossett 2013). In either case, the discourses and pragmatics of care were characterised by an 'overt metaphysical/religious pluralism' (Connolly 1999, 185) that residents experimented with and appropriated to create hybrid and increasingly complex interplays between religious, spiritual, and therapeutic practices.

Second, despite staff tending to frame motivation in relation to broader faith and secular identification, there were several discursive prompts that aided the construction of shared ethical motivation. For example, several staff – religious and non-religious – cited how personal histories of alcohol and drug use led them to empathetic identification with service-users:

> [I wanted to] help people who are going through the same problems as I did.
> (Mark, rehab counsellor 3/8/10)

> [Y]ou never just disclose your own history, I can empathise a lot with what some of the lads have been through because there are certain things in life

I've done, choices I've made, so I'm able to see how Christ had come to me and gave me hope and freedom through stuff.

(Stephen, centre manager 12/1/10)

This opened out a convergence space that partly dissolved divisive identification of religious and secular motivation. Equally, staff who self-identified as non-religious saw the decision to work for a Christian charity as largely pragmatic, reflecting more of an acceptance of TSA as a practical device 'just to make a difference' (Neil, head of outreach 12/9/10) rather than any systematic approval of religious belief *per se*. Respondents affirmed that the Christian ethos of the TSA was not a barrier to their participation, pointing to the synergy between the outworkings of faith-based and secular ethics:

[T]he values are pretty similar to my own, you don't have to be a Christian to be loving.

(Sharon, rehab counsellor 5/8/10)

I suppose we are all here because we're human, we [pause] care, the only thing we are here to do is help people get back on their own feet.

(Dave, senior detox nurse 16/8/10)

[I]t doesn't really matter what you believe, there's such a mixture of us anyway. I have my own reasons for working here, they have theirs. Christian, Muslim, Buddhist, or if you're simply just a human caring for another human, we all have something that drives you to do this work. I came here not because it's particularly Christian but because I could easily agree with its TSA ethos – caring for people, yep, helping the whole person, yep ... in my opinion, it's [the centre] not overly religious in the end of the day, we're not evangelical.

(Tasha, rehab counsellor 16/8/10)

Different religious or humanist motivations were seen to discursively frame phenomenological and embodied responses to care for people struggling to overcome substance misuse and addiction. Each of the interviewees above articulated a shared intuition – to care, to love – a sensibility that led actors to commit doing this sort of work.

Third, we suggest that this mutual recognition of the ethical capacity of both religious and non-religious motivation helped co-generate hopeful sensibilities and ways of working that reflexively transformed religious and secular mentalities. To illustrate this, we draw on the example of the receptionist, Joy, who embodied intentional and routinised performances that helped to create representational and emotional-affective landscapes conducive to rapprochement. Joy's daily routine involved operating the security door, dealing with all enquiries, collecting and giving residents keys, and organising appointments for residents and staff. Her jolly and calm demeanour was often commented on by residents and staff alike. She had a knack for 'picking people up', interspersing humour, small talk, and remembering all the residents' names. Joy would dispel aggravated situations when residents were 'kicking off' by simply getting alongside people and taking time to listen. The way she performed her job communicated certain affective and visceral messages to others. For instance, despite sitting behind a glass-fronted reception desk, when residents came with queries Joy would leave her seat to lean over the counter to offer an embodied gesture of hospitality and individual importance to clients. Sometimes she would leave the locked reception office to embrace residents in the lobby, or simply go and join one or two of the men having a cigarette outside the building, thereby revising some of the staff–resident hierarchies that can characterise professionalised welfare spaces.

Joy openly ascribed the way she performed her job as 'part and parcel' of her evangelical Christian belief to 'show God's love in practical ways that are meaningful to the men' (Joy, receptionist 15/8/10). Equally, Joy articulated experiences of being 'empowered by God to love that one person' and feeling 'God was working through her' (Joy, receptionist 15/8/10). This theological belief was seen to attribute agency to the divine in ways that led the believer to enact particular embodiments of caritas and agape, suggesting the need to recognise the distinct psychogeographies at work for people of faith (Dewsbury and Cloke 2009, 696). The claim here is not that Christian motivation somehow produces a stronger display of caring and warmth, more than, say, secular, humanist, or humanitarian motivation. Rather it acknowledges that different ethical precepts performatively elicit distinct affective registers. In this example of Christian belief in the immanent and invisible powers – which in this case referred to Joy's belief in the kingdom of God – this means that 'certain things happen that would not otherwise – certain affects are produced that make people experience very real and specific feelings' (Dewsbury and Cloke 2009, 696). Here Joy's habitual performance, driven in part by a theo-ethics of agape, came to shape the emotional tonality of the reception space which evidently prompted similar performances of care from secular and other Christian staff. Certainly, when Joy was not on shift the atmosphere in reception did not convey the same affective lines of hospitality. We argue that it is this distinct constitution of an emotional-affective landscape of care that allowed actors – religious or nonreligious – to recognise the salience of beliefs-in-action. In this example, the charismatic personality of one individual, combined with her generosity and hopefulness informed by Christian theo-ethics, helped shape an affective texture that spilled over into the sensibilities and practices of religious and non-religious actors.

Hopeful re-enchantment

In the examples above, we see that conceptualisations of faith should not neglect non-religious forms of faith or enchantment in narratives of unconditionality, universality, and humanism. Faith can be taken to define an embodied sense of religious belief, but it can also suggest some form of secular belief or commitment, or a completely different form of fidelity to an idea (see Critchley 2012; Holloway 2013). The case of TSA's Hope House exemplifies actual existing postsecularity at work whereby religious and non-religious motivated practices of care come together and so recognise shared ethical precepts of love, hope, and compassion. According to Critchley (2012), faith is a commitment, a proclamation of fidelity to an ethical demand that enacts a new form of subjectivity – an emptying out of the self towards an 'other'. Faith is not necessarily related to transcendental belief but to an event that is shared by agnostics, atheists, and theists alike. Here Caputo's (2001) reformulation of the distinction between religious and secular persons usefully illuminates what he argues is the inherent religious characteristic of practices of going-beyond-the-self, or love-as-excess. In his treatise on 'religion without religion', Caputo indirectly draws on Kierkegaardian existentialism to open up a kind of endless substitutability and translatability between 'love' and 'God'. Rather than distinguishing between religious and non-religious people, he argues, it is better to speak of the religious in people – where a leap of love into the hyper-real leads to a transformative commitment.

In Hope House, we see a fluid constellation emerging among religious and non-religious actors, centring on an experiential and performative stance towards faith-in-practice beyond dogmatism. It is galvanised by mutual recognition of the subversive capacity of religious and secular voices exercising their spiritual capital in practice; but is rooted in a phenomenological response to need that reconfigures desire towards hopeful transformation. Religious, secular, and humanist motivations take on a shared 'belief in the impossible' becoming possible – expressions of love steeped in an obstinate hope for transformation, an 'impossible' belief that someone or something can change when there is so little sign of it (see Caputo 2001). Caputo (2001, 13) marks out this distinction between the hope of the mediocre fellow – 'the sanguinity that comes when the odds are on our side' and the more self-surpassing passion 'hope against hope', as St Paul says (Romans 4:18). He writes:

> [I]t is no great feat, after all, to love the loveable, to love our friends and those who tell us we are wonderful; but to love the unloveable, to love those who do not love us, to love our enemies – that is love. That is impossible, *the* impossible, which is why we love it all the more.
>
> (Caputo 2001, 13)

This faith-in-the-impossible, or venture into the hyper-real, is the crux of a shared faith that sustains religious and secular rapprochement. In Hope House, this was clearly seen in the offer of second, third, and fourth chances to people who had violated the rules of the centre, or in the choice of staff to work with 'difficult' residents who had threatened other staff and residents. These hopeful sensibilities were seen in more ordinary practices of 'going-beyond-the-self', where religious and secular staff voluntarily stayed on 'after hours' with residents and befriended residents on the detoxification unit struggling with withdrawal and loneliness.

Crossover narratives and rapprochement as liminality

What is evident in Hope House is the possibility for postsecular rapprochement to generate thirdspaces where previously non-negotiable views are deterritorialised and new lines of flight and hybrid forms of belief across religious and non-religious identities emerge. In the liminal space that emerges when religious and secular beliefs are negotiated, it has been argued that 'new values may well be formed as part of a valuable poststructural reterritorialising of the faith-in-practice of postsecularism' (Cloke and Beaumont 2013, 47). To illustrate this, we present two broad reformulations of secular and religious identity and belief among staff and residents.

First, staff identifying as secularist who had fixed views on the detachment of religion from public service delivery came to recognise shared intuitions and convergence points across religious, humanist, and secular sensibilities. In some cases, there was a reluctant appreciation of the role that religious theo-ethics of caritas played, when combined with the Salvation Army ethos, in fashioning a 'caring' centre that 'brings the best out of the men' (interview with Richard 3/8/10). This it does by providing an 'atmosphere of understanding and acceptance' (interview with Tasha 11/8/10). The pluralistic therapeutic discourses accompanying person-centred care led atheist staff to cultivate a receptive generosity, or what Connolly (1999) calls critical responsiveness, in relation to the positive role that religious and spiritual beliefs might play in the lives of residents. Equally, for residents, there was an appreciation of the openness towards spirituality in treatment and the willingness to respect differences in worldview among staff and residents:

> They don't shove Christianity down your throat. And they allow you the space to develop whatever your belief system is, or wants to be, or needs to be or whatever.
>
> (Ali, rehab resident 20/7/10)

> Andrew's very good as well, he just says what he believes and allows you to say what you believe, and … cause the answer is that none of us really know, it's a matter of faith isn't it?
>
> (Colin, detox resident 20/7/10)

These responses were combined with an acknowledgement that Hope House engaged values of non-judgementalism and tolerance of difference compared with other faith-based and secular rehabilitation centres they had visited. Religious monologues or preaching were deemed 'out-of-place' in the centre, which according to several residents permitted a more open dialogue between different standpoints, and in some cases led to new hybrid composites of different religious, scientific, and spiritual resources in residents' narratives of recovery.

Second, the entanglement of disparate metaphysical/religious identities and expressions fostered a critical responsiveness among Christian staff – prompting reflection on the limitations of Christian caritas and the practices of unwanted proselytisation (Coles 1997). Working across faith and secular boundaries led to a greater affirmation of the power that 'other' religious, secular, and humanist beliefs play in motivating a performative excess of care. In the interviews, staff described moments when the willingness of New Age staff 'to do stuff which isn't part of their job description' (Molly, former doctor 3/9/10), as part of their desire to show compassion to residents, led to a reconsideration among evangelicals of the legitimacy of New Age spirituality.

By the same token, the overt metaphysical/religious plurality meant religious and secular participants had to negotiate and translate their aspirations and deeply held convictions with each other. Sometimes people had to leave because they felt too compromised. Others saw it as a form of liberation in which their own tradition becomes more alive and more credible. Resettlement worker Paul's experience working in the centre, for instance, gave him a heightened sensitivity to the erroneous identity-politics of conservative evangelicalism, which presented his role as 'saving souls' and 'bringing people to church' (interview 15/8/10). Working across faith-secular lines for him prompted a shift towards a theology of *Missio Dei* – recognising 'Jesus is already at work in people's lives regardless, [and] my job is to assist in that' (interview 15/8/10). Accordingly, Paul and others embraced theologies and practices that affirmed the revelation of God in expressions of love shown by religious and secular actors:

> Wherever there is truth there is Jesus ... wherever there is kindness, there is God. God doesn't just use Christians you know.
>
> (Joy, receptionist 15/8/10)

This example seems to resonate with Caputo's translatability between 'love' and 'God'. Yet for Joy, an evangelical Christian, this theology of immanence became a question about the inherited systematic theology in her church community. Accordingly, the deterritorialisation of theological belief is a fragmented and unfinished process, where rapprochement opens out the possibility of deconstructing the intuitive embodiments of religion and its reliance on particular institutional codes. Furthermore, the move towards faith-by-practice was seen by some to offer experimental space to 'reveal the true kernel of radical openness [of the church], but also transformative alterity that lies at its heart' (Baker 2013, 10). This can be clearly seen in Paul's attempt to experiment with more dialogical expressions of faith-sharing and bottom-up, or rhizomatic, theologies that seek to rediscover 'how to be church in the community – with homeless people – outside the four walls of a [church] building' (Paul, resettlement officer 15/8/10).

These specific movements in the politics of becoming were accompanied by a revalorisation of Christian virtue ethics that prioritised a faith-through-praxis over propositional belief, which enabled a greater spirit of ecumenism across the religious–secular divide. Asked whether Hope House was a Christian programme, several evangelical Christian interviewees downplayed the place of mandatory, unwanted proselytisation, emphasising instead the importance of learning to respond consistently in line with the character of Christ (see Hauerwas *et al.* 2010):

> We can't save everybody, we can't convert everybody. We need to accept that … But I think it's important here in this kind of situation because I believe the character that you put on of Christ is the character the people here that are resident need.
>
> (Esther, trainee manager 27/8/10)

> It doesn't matter if I'm doing that [compassion] as an officer or if Joy is doing that as a Christian [then] that is a Christian programme to me … there's people from the Salvation Army, there's people from other churches and then there's non-believers, there's a mixture and I think people aren't afraid to be themselves and I think that is quite an amazing thing to see…
>
> (Esther, trainee manager 27/8/10)

From these short reflections, we want to suggest the possibility for postsecular rapprochement to generate thirdspaces where previously non-negotiable views are deterritorialised, and new lines of flight and hybrid forms of belief across religious and non-religious identities emerge.

Rapprochement as contested

Rapprochement emerges from the negotiation of existing secular and religious divisions – a process that by nature holds the possibility of entrenching existing religious and secular identity boundaries as much as fostering new relations of mutual translation across secular/religious boundaries. Indeed, rapprochement between secular and religious belief was seen to negotiate more longstanding discursive codes that in some cases worked to entrench strict religious and secular identities. For some individuals, there was a secularist aversion to the presence of any 'religious talk' in the centre, while for others the negotiation of religious and secular voices was perceived to be part of the 'secularising' tide within TSA diluting its 'Christian' character.

Crossover narratives also need to be situated in relation to the social and organisational context in which they emerge. On an organisational level, rapprochement might be seen in the ethic of hospitality on the part of a Christian FBO shown to other monotheistic and Buddhist faiths, particularly with regard to efforts made to encourage residents to harness the emotional and spiritual support from their own faith community. Certainly, any criticism of non-Christian faiths was seen immediately to be 'out of place' within Hope House. However, the Salvationist management of Hope House upheld a boundary of tolerance in terms of what lines were drawn between acceptable and inappropriate expressions of religious and spiritual belief. While the practices of 'mainstream' religions were tolerated, even encouraged, some New Age and Pagan members of staff at times were discouraged from sharing their own spirituality with clients. For example, Katie, one of the detox nurses was given 'a slap on the wrist' when she gave tarot card readings to several men on the Drug Programme (Katie, interview 12/8/10). Her request for senior management's permission to offer rock and crystal therapy to residents who had expressed an interest in New Age spirituality was declined, citing how this contravenes the ethos of the Salvation Army.

This raises the question of the power dynamics underpinning cases of post-secular partnership and how disagreement between disparate positionalities is managed. At least two strategies can be observed at work. First, senior man-agement used the central TSA image to jettison some proposals as 'out of place' in a Christian organisation, while circumscribing acceptable discursive practices through the rules, staff performance reviews, and 'corridor talk' that regulate the space. In the interviews, several respondents recalled instances where senior staff dismissed the use of alternative spirituality in the treatment programme by parodying the need for 'auditable interventions', implying how that would look in the eyes of commissioners and funding bodies. Second, heated conversations between staff members of disparate beliefs were inform-ally moderated by the presence of staff holding alternate viewpoints – coming from different denominations or intensities or identifying as non-religious – as well as being curtailed if there was a sense the disagreement was giving a 'bad impression to clients'. This highlights the ambiguous and contested nature of postsecular spaces, especially the asymmetrical power relations that structure the possibility for rapprochement.

Nevertheless, the example of Hope House represents a case where postsecu-larity is seen to be primarily produced through the affective atmospheres and liminal spaces of sociality that generate an embodied and cognitive recognition of common ethical capacity and desire. Religious and secular collaboration is sustained partly through ephemeral, visceral embodiments of shared faith, but also through the construction of crossover narratives and devices capable of holding together the combined discourses and praxis of secular and religious actors.

Postsecularity and emergency food provision: the case of Levington food bank

Food banking has received considerable criticism by academics on a number of grounds: for cementing processes of welfare reform (as food parcels replace hardship funds); helping to legitimate damaging constructions of the dangers of 'welfare dependency' and a paternalistic stance towards 'the poor'; and depoliti-cising problems of food poverty (by apparently meeting the need for emergency food without confronting the reasons for that need) (see Lambie-Mumford 2017; Garthwaite 2016; Williams *et al.* 2016 for a review). It is a deeply contested ethical terrain among different anti-poverty organisations, some of whom are suspicious of the faith-based and entrepreneurial character of leading food poverty organisations. The growing corporatisation of food charity (Fisher 2017), now seen in the UK with the closer partnership between corporate philanthropy, agribusiness, and two major food aid providers – FareShare and the Trussell Trust – means food banking represents a deeply ambiguous context to examine emergent postsecularity.

Yet postsecularity is evident in a variety of spaces related to contemporary food aid provision in the UK. Three have been identified to date. First, on a *national* level, foodbank organisations such as the Christian charity, the Trussell Trust, have been prominent in shaping the political and public debate about food poverty. In absence of more comprehensive measurements of food insecurity, the Trussell Trust's early role in providing evidence about the scale and causes of food insecurity and poverty in the UK opened out a highly politicised space in which pragmatic collaborations across ideological and religious difference have emerged. One of these movements has been the End Hunger UK campaign – a diverse coalition of national charities, faith groups, frontline organisations, academics, and individuals working to tackle food poverty in the UK (http:// endhungeruk.org). It was launched in 2016 aiming to build a 'food justice movement' that is 'united in the belief that to really tackle the root causes of household food insecurity we require a concerted effort by the UK and devolved governments' (End Hunger UK 2017, 4). In doing so, members have demonstrated a willingness to put aside ideological and religious differences among organisations, recognising that it is only through partnership it will be able to mobilise the 'thousands of people engaged in tackling food poverty in its many forms [and] build a powerful movement for change' (End Hunger UK 2017, 3). Faith groups have also been leading representatives in several parliamentary inquiries and local alliances on tackling food poverty such as *Feeding Britain*. This is in part a reflection of the landscape of emergency food provision, with the majority of foodbanks linked to churches, but increasingly alliances on this issue include tenant associations, mosques, schools, GP surgeries (May *et al.* 2017). It has also been shaped by the renewed public acceptance of (left-leaning) religious voices in the political debate over hunger, austerity, and welfare. For instance, in April 2014, 40 Anglican bishops and 600 church leaders called on all parties to tackle the causes of food poverty, invoking the moral authority of Christian tradition to attack government policy of sanctions that normalise destitution. This led to an immediate backlash by conservative MPs, including the then Prime Minister David Cameron, who criticised the Bishops' Letter as religious infringement in public affairs and defended welfare reform as a 'moral mission'. The role of religious voices in publicly contesting government policy has fed into the mobilisation of 'communicative publics' – both nationally and city-wide – made up of a diverse set of actors including antipoverty campaigners, food bank volunteers, researchers, journalists, tweeters, politicians active in challenging dominant narratives around welfare 'reform', welfare recipients, and engaging in political campaigning. Spaces of care such as foodbanks become re-scripted as ways of expressing anti-austerity sentiment that facilitate wider ethical–political alliances across voluntary organisations and campaigning groups.

Second, in the *localised* provision of food aid, there are numerous examples of explicit rapprochement across lines of religious difference. Sufra NW London, for instance, was set up in 2013 to initially provide emergency food for people experiencing poverty in London, and has evolved to provide food growing projects, cooking skills, employment courses, and vocational taster courses. Based in St. Raphael's Estate, a disadvantaged neighbourhood in Brent, Sufra exemplifies postsecular rapprochement initiated in and through the Muslim voluntary sector. According to its founder Mohammed Mamdani, 50 per cent of funds and resources come from the Muslim community, with the rest coming from other faith communities including the local Catholic Church and the Jewish community. As an organisation it does not neatly assign to the label 'Muslim-led faith-based organisation', but instead deliberately welcomes people of faith and nonfaith backgrounds to create a 'sustainable common purpose' and is therefore better understood as an interfaith organisation with multi-faith volunteers and staff. In the early days of the project, Mohammed Mamdani explained he faced challenges from people who assumed that his project would not work with LGBT people. Not only is a referring agency for his centre a local LGBT group, but he said that his centre aspired to be an organisation where people of different faiths and secular backgrounds could 'take part in social action together, fundraise together, and share resources together' to create what he calls a 'sustainable common purpose'. This is borne out by the fact that 90 per cent of people referred to Sufra are non-Muslims, and the project attracts volunteers from all faiths and none. In performing these roles, faith groups are pivotal hubs and curators of new expressions of postsecular citizenship and a deeper form of politics based on a renewed sense of hope and resilience, rather than the anti-politics of despair.

Third, foodbank spaces can represent *sites of encounter and politicisation* which reconfigure the ethical and political subjectivity of actors involved. Rather than dismissing foodbanks as simply a mechanism that institutionalises charitable assistance in replacement for case entitlement, or a symbolic gesture that depoliticises and placates energies for addressing the structural processes underpinning poverty and food insecurity; Williams *et al.* (2014) have suggested voluntary spaces of care such as foodbanks present a 'practical device' through which citizens demonstrate a capacity to set aside moral, religious, or ideological differences in order to respond ethically to immediate social need and embody a politics that refuses to accept the inevitability of austerity – combining national and local campaigning as well as local advocacy and practical care for individuals.

To unpack these broad dynamics of postsecularity further, we focus on the example of Levington FB (hereafter LFB, see Williams *et al.* 2016) to showcase the ways in which food bank spaces constitute liminal spaces of encounter and politicisation – via the creation of 'micro-publics' among volunteers of different social and political backgrounds, and the potentiality for food bank volunteering to reconfigure ethical and political subjectivities.

Background

LFB is based in relatively affluent city in Southern England. Part of the Trussell Trust foodbank network, it opened in 2008 through a partnership between a small group of churches in the city. In 2016/2017, LFB provided 4,668 food parcels to people experiencing food poverty in the city. At the time of the research (Williams *et al.* 2016), the foodbank had over 100 volunteers and employed two part-time staff. The volunteers were predominantly white, middle class, affiliated to congregations, with the majority aged between 50–70 and retired from paid work, with a smaller number of students and young professionals. There were slightly more female than male volunteers, and unlike some foodbanks in our research, few had used the foodbank themselves. LFB is an example of postsecular rapprochement that originates from within the organisational apparatus of Christianity. The food bank operates out of a Methodist church in the centre of the city; team leaders and staff members are required to be Christians; there is a short time of prayer before opening times led by the team leader; some volunteers offer prayer but only if requested by clients, and adjacent to the space where clients sit and drink refreshments, leaflets advertising the Christian basis of the foodbank can be found alongside multi-agency information and advice on debt advice, counselling, and homelessness, amongst other topics. Despite these clear Christian roots, in recent years, the food bank has adopted a 'postsecular' stance (Cloke and Beaumont 2013); welcoming increasing numbers of volunteers from a range of religious faiths and none, including those who simply want to 'do something to help' after reading articles about food banks in the national and local press, and those whose primary motivation is ideological – embodying solidarity in the face of welfare austerity. Together with the 'apolitical' marketing of the Trussell Trust franchise, which has studiously avoided alignment with any specific political party, volunteers expressed an assortment of ethical, theo-ethical, and political standpoints on issues of welfare, poverty, and austerity. These range from politically conservative standpoints on welfare dependency, deservingness, and rationing to notions of mutualism, solidarity, and reciprocity. What is significant here is the rapprochement across volunteers of different religious and ideological motivation, often putting aside oppositional moral and political standpoints to 'do something' about food poverty.

Space of encounter and politicisation

Within the space of rapprochement in LFB, foodbank volunteering can also be generative of potential spaces of encounter through which existing ethical and political attitudes, beliefs, and identities can be reinforced, reworked, or challenged (Valentine 2008; Lawson and Elwood 2013). Volunteers shared experiences of becoming 'sensitised' to people's stories – an experience that led to a sharper perception of broader structural issues, including, zero-hour contracts/underemployment, delays and cuts to welfare payments, and the impact of specific welfare reforms, such as the 'Bedroom Tax'. While for some this sensitisation clearly mapped on to existing political proclivities, for others, the experience of working in the food bank came to disrupt received views on poverty:

> I have become more aware of how our government seems to push through measures without really thinking about the consequences and the damaging effect on some people. In particular the so called 'bedroom tax' ... or the way people can be penalised for trying to find employment. There is a lack of real understanding and compassion.
>
> (Esther, 50s, volunteer, January 2014)

> I have certainly become more aware of how some people are really struggling. I have become more aware of the short-comings of the benefit system ... [and] of how redundancy, accidents and poor health can have a 'knock on effect' and almost destroy people.
>
> (Abby, 60s, volunteer, March 2014)

> I think my approach was quite patronising when I first started. I had an idea that it's a good thing to do, to help people in need. Which is fine. I wasn't very politically aware. And I think gradually it made me look at the bigger picture ... at some of the causes. Whereas initially I was motivated by a sense of just grace, I suddenly became more integrated with a sense of injustice and feeling that more needed to be done to address the causes of that. And actually, just the experiences of sitting down and talking to people had an effect on what I did with the rest of my life.
>
> (Lydia, 20s, volunteer, March 2013)

As she recounted her experiences at the food bank, Lydia also charted a significant shift in her understandings of poverty and of the most appropriate response to problems of poverty: from one rooted in Christian conservative attitudes (privileging an avowedly 'apolitical' and 'patronising' understanding of charitable giving to the 'needy') to more explicit, and explicitly politicised, expression of faith that embraced the need to 'do something' at a structural as well as a local level. This emergent framing shaped her understanding and response to food poverty. It not only considerably changed how she related to clients, but also helped her to effect broader changes to her life as she took up a part-time paid position as an administrator at the food bank and at a local homeless charity. Her work at the food bank led to broader changes which 'spilled out' into other areas of her life. The work of other volunteers also often 'spilled out' beyond the food bank itself; with volunteers taking food back to client's homes (by hand or by car) if people were unable to carry the food parcels themselves, for example, or – in rare cases – with volunteers purchasing toasters and kettles at their own expense for clients who did not have the means to cook the food provided.

In doing so, volunteers of different political persuasions found that identification with the lives of foodbank clients, and by association the politicised debate surrounding food banking, led to a deeper emotional connection with the unjust and complex circumstances experienced by those whose lifeworlds are often caricatured on both sides of the political spectrum. Through this, previously held positions of Christian conservatism that separated charitable food provision from politics came to be questioned and integrated with a deeper understanding of the lived effects of welfare 'reforms'. Most significantly, some volunteers who had previously identified as conservative or disinterested in politics shared that listening to client stories had led to a sharper emotional response to the failings of government policy and a greater willingness to engage in various forms of 'anti-poverty' activism, including lobbying MPs and joining local anti-austerity groups in the city. Crucially, a number of volunteers became catalysts in their wider social networks: engaging and challenging discourses surrounding poverty and deservingness, alongside working to recruit other volunteers and supporters willing to speak about issues of food poverty among their own social networks. Lydia, for example, commented:

> I think I was also quite shocked by some of the attitudes within the church as well, a lot of the language of 'their responsibility, they are in this situation, they have created it,' and I didn't see that that matched up. What I could see from my reading was being reinforced. I just wanted to help people see that.

This example illustrates the ways in which one person's experience in the foodbank led to a critique of a particular set of beliefs which in turn opened up discussion in the church about its responsibility to get involved locally, by supporting the foodbank, as well as its involvement, more widely, with regard to, for example, political campaigning. Inadvertently, foodbanks present opportunities for encounter between people of different backgrounds, ideologies, religious belief, social status, and financial wealth. These encounters potentially facilitate dialogical learning and reflexive engagement, where the process of listening to marginalised voices leads to ethical stances of receptive generosity and reconfiguration of desire. For volunteers like Lydia, this was characterised by a renewed sensitivity over the need to show love unconditionally and make space within the self for 'otherness' without judgementalism: refusing to pre-judge the identities, histories, and experiences of people 'unlike oneself' and instead enter into a reflexive and mutual engagement that allows the voice of the 'other' to speak back. In this way, liminal spaces of encounter in the space of the foodbank came to generate new, and rework existing, ethical and political sensibilities that contested the myths and stereotypes surrounding poverty, deservingness, and welfare dependency. Perhaps one of the more hopeful murmurings found in food bank themselves, therefore, is their role in revitalising the discernment of (in)justice in congregational and social networks, and energising faith communities – especially those that might be more politically and theologically conservative – to act ethically and politically.

What we see in some food banks then is an inherent potentiality for encounter and transformation – allowing the self to venture out of its ever-narrowing social networks and enter dialogical engagement with people 'unlike them' – a process that can reconfigure an individual's views and expressions of politics and religion. Alongside their primary role in providing support and sustenance for people experiencing food insecurity, often due to the violence of neoliberal austerity (precarious low paid work, sanctions, benefit delays/changes, restrictive eligibility), foodbank spaces also open up possibilities for the (un)learning of ethics and the contestation of hegemonic political constructions of poverty. This is at the heart of the argument for post-disenchantment: dwelling in spaces that help discern, name, and secede from the enchantments of late capitalism, its seductive telos of 'success' (measured by popularity, achievement, wealth, and power) and normalised affective dispositions that enculturate the subject-citizen of possessive individualism. Mundane participation in spaces of care and welfare can solicit encounters that can lead to the cultivation of values that cannot be easily commodified by the market-state, and also generate energies that escape the trappings of institutional religion. The phenomenology of need, as argued by Cloke *et al.* (2017, 708), can be an incubator for civic virtues such as altruism, generosity, and solidarity which not only offer

> competing conceptions of appropriate social practice … [but also] potential for meaningful encounters between people of different social positions and, as a consequence, some political and ethical re-evaluation of what constitutes the common good, and how it might be cared for.

Understood in this way, foodbank spaces are re-read as sites that can shed light on and challenge the normalcy of capitalist enchantments; as well as generating dispositions and senses of ethical responsibility that recognise and act upon the mutual recognition of in-commonness. This involves moving beyond relatively apolitical expressions of charitable endeavour towards experimental spaces that develop more reciprocal, mutualist, and conscientised forms of food sharing and activism. The call for post-disenchantment (Rose 2017) also speaks to the need to secede from right-leaning religious enchantments – which encourage eschatological postures of disengagement from the 'worldly' affairs of politics and social justice, sacralise exclusivist and exclusionary theologies, and limit receptive generosity to a particular social and spatial extension. It is telling that the experience among some foodbank volunteers resonated with this disenchantment, or 'deconversion' (Bielo 2013), from right-leaning religious beliefs, with some volunteers disclosing feelings of being 'estranged' from elements of their church culture and its theological priorities that they once held dear and took as self-evident.

Partiality and the limits to postsecularity

What can this particular example tell us about the spatial formation of subjectivities associated with geographies of postsecularity? First, it underlines the *partiality* of rapprochement. Not all participants will enter into the thirdspace of receptive generosity. Some individuals will continue within fixed frames of reference – influenced by theological commitments and political representations of welfare recipients – that shape rather closed and unreceptive stances towards people making use of the foodbank. For some individuals, the politicisation of food banks became too difficult to square with their own conservative values and stopped volunteering entirely (Williams *et al.* 2016). Equally, for some volunteers, expressions of charity retained a degree of expectation that 'clients' should conform to a particular set of norms and behaviours (not being fussy, displays of gratitude, and so on). Here, motivation was narrated through a form of Christian conservatism which was susceptible to condescending language of 'helping the needy' and associated with methods and theologies of 'rescuing the poor' – bringing 'them' into the spaces, practices, and identities that are more comfortably familiar to 'us'. For other volunteers, addressing food poverty was simply considered the alleged 'non-political' domain of Christian charity which in various guises remained wedded to right-leaning discourses of the dangers of 'dependency' and only helping 'those in genuine need'. If this was the only political discourse constructed in food banks, then, such spaces of welfare can be rightfully criticised as cultivating a 'quiet sense of the ordinary' (Ehrkamp and Nagel 2014) – a gesture of charity that at best eschews structural questions of economic injustice, and at worst, works within class-based and racialised imaginations that reinforce the very exclusions they seek to alleviate. Put simply then, the journeying in and out of the real and imagined space of rapprochement is strewn with difficulties, and it is possible for hardened discourses of poverty alongside religious and secular fundamentalisms to remain dormant, or unarticulated, in partnership.

Second, relational and physical proximity to 'poverty' is no guarantor for ethical transformation, and there are crucial *temporal and spatial limits* for those who engage with practices of receptive generosity and allow the voice of the other to shape one's own ethical and political subjectivity. Foodbank encounters are usually temporal, suggesting that relationships of receptive generosity can be short-lived. It also concerns the contingent character of the foodbank itself. The diverse motivations and political attitudes among volunteers, alongside the way volunteers perform their role, can produce a diverse set of experiences for clients, ranging from care/oppression, generosity/stigma, and acceptance/shame. The overt charitable roles of giver and receiver can potentially override and close down opportunities for meaningful engagement. In addition, the manner in which clients narrate their story plays a key role in shaping volunteers' perceptions and attitudes towards issues of 'food poverty'. Clients often worked to navigate the charitable social identities and roles – perhaps inadvertently – ascribed to them. For instance, some clients explicitly challenged dominant media stereotypes of foodbank users as somehow responsible for their own plight (through bad decisions, behaviours, and motivations) or as victims of personal misfortune (benefit sanctions and delays); and instead, sought to give volunteers insight into the very real connections between larger scale and systemic processes that produce heightened vulnerability, and the person using a foodbank. Nevertheless, the transient and fast-paced service environment of the foodbank, combined with the scripted identities of 'giver' and 'receiver' presented barriers to long-term reflexive engagement with others.

Third, ethical conscientisation will often be non-linear in character. There is no straightforward transition from an apolitical to an overtly political position, or from self-identifying charitable endeavour to receptive generosity. Subjectivities of postsecularity slide in and out of the seductions of 'moral selving' (Allahyari 2000), and the short-term character of these spaces of encounter might allow moral distancing to creep in whereby client stories are later retranslated through the dominant lexicon of dependency and assumed culpability.

Fourth, ethical stances of postsecularity can be *purposeful*, emerging in the foodbank space despite scripted roles and the materiality of the service environment that reinforce the power dynamic of giver and receiver. Stances of receptive generosity are performative and can bubble up in unexpected moments of sociality, laughter, sharing of vulnerability between foodbank volunteers and users. In the space of the waiting room, the café area, or through playful interruption of children, affective atmospheres within foodbanks are highly situational on context and are differently experienced according to individual positionality and biography, especially prior experience of accessing support from voluntary and statutory service environments. To acknowledge the ways in which food bank spaces are 'peopled' reveals the ethical agency that volunteers, staff, and people using the foodbank bring to that space. While foodbanks are embroiled in wider governmentalities of shame and stigma (Tyler 2017), we suggest this might not always be the dominant affective atmosphere at work and experienced in foodbank spaces. Instead, foodbank spaces more closely resemble a complex amalgam of care and stigma, sociality and obligatory gratitude, judgement and acceptance.

Fifth, the possibilities for hopeful re-enchantment and politicisation will remain inconsistent and contradictory if food bank organisations, volunteers, and clients do not challenge the enrolment of food bank technologies into dominant discourses of welfare dependency that work to embed the politics of 'othering' in postwelfare societies. In this example, emergent postsecularity seems to operate within demarcated limits whilst being *entangled* within wider unjust relationships. Practices of receptive generosity will be at best partial and contradictory if they do not acknowledge the need to address the complicities inherent to many charitable projects which buttress systems of exclusion and politics of 'othering'. The Trussell Trust, for example, continues to use a voucher and referral system that draws upon, and risks further legitimating, a neoliberalisation of welfare based around evermore-strict distinctions between the 'deserving' and 'undeserving' (Williams *et al.* 2016). The modus operandi of many foodbanks has been criticised for upholding, perhaps inadvertently, imaginaries of dependency and the necessity of rationing that further the legitimation of replacing cash entitlement with payments in kind (ibid.). Indeed, the growing move towards corporate partnership with Tesco and Asda suggests a disconnect between the emergence of politicised subjectivity at a local level and structural dilemma of institutionalising food charity and issues of exploitative labour and environmental practices within the food industry. Asda's £20 million partnership with FareShare and the Trussell Trust in 2018 does little to address the 120,000 low-paid staff in Asda who receive working tax credits and other public subsides totalling £221,337,000 because their paid employment does not provide enough to live on (Cox 2015).

4.4 Conclusion: divergent possibilities of collusion and subversion

What has become clear from the research detailed in this chapter is the variegated geographies of postsecularity and the inherent potentiality for *both* subversion and collusion within neoliberal formulations. Ethical subjectivities inspired by practices of receptive generosity, rapprochement, and hopeful enchantment clearly are not immune from ambivalent entanglement in neoliberal logics, exclusionary power dynamics, and the perpetuation of gendered and racialised injustices. This raises the question as to whether the allegedly 'progressive' promise of postsecularity is simply a chimera; especially given that the examples above could be dismissed as being co-opted into reinforcing the neoliberal logics of charitable endeavour.

How, then, can postsecularity be considered as reworking or resisting the dominant political-economic order? We want to acknowledge these ambivalences but also suggest the choice between co-option *or* resistance is perhaps not the most helpful framing device. Instead, we have advocated for an analysis of the possibility of 'both/and' in the messy middle on two grounds. First, we have highlighted the performative potential of subversive ethics in reworking the enactment of neoliberal logics, cultures, and relationships. This is seen in the direct resistance, protest, and advocacy work of organisations such as Z2K. However, even in the contracted arena of service delivery, TSA Hope House sought to rework the intended technologies and subjectivities supposedly normalised in the regulatory frameworks of neoliberal governmentalities – namely individual compliance and evictions for violation of rules. Some might dismiss this as the latest mutation of neoliberal governmentality – the use of care, empathetic authority, unpaid labour – as tools to make 'irresponsible' subjects governable through self-regulation. To do so, however, misses the inherent potential for a quiet politics (Askins 2015) and implicit activism (Horton and Kraftl 2009) to emerge *inside* neoliberal spaces of governance in ways that subvert the enactment of uncaring logics of deservingness and just rewards that divide, label, and stir up suspicion and resentment between people. Indeed, through the case-study of Hope House, we have highlighted the ethical and political spaces created through rapprochement between religious and secular ethics, specifically the ways in which postsecularity is galvanised by a desire to counter, even resist, the dominant rationalities of postwelfare politics.

Second, we do not want to contest the veracity of established critiques of voluntary welfare spaces in neoliberal governance; rather we consider postsecularity as a supplementary narrative that allows a critical and empirically sensitive re-reading of these spaces, attuned for difference not domination (Gibson-Graham 2006; May and Cloke 2014). In this way, religious and secular collaboration in spaces of welfare can be re-read as liminal and transitional spaces of encounter, deliberation, and transformation. In this conceptualisation emphasis shifts to an analysis of precorporation and the politics of ethics itself. As the foodbank case-study demonstrated, in these spaces there is nascent potentiality for the unlearning of ethico-political attitudes – stances of cynicism, individualism, materialism – and the curation of new ethical and political subjectivities. The claim is not one of exclusivity: we are not claiming that religion, or spaces of postsecularity more specifically, are somehow guaranteed to be successful in bringing about these transformations. Rather, there is a potential within these spaces, a vibrant openness that sparks off intuitive and dialogical negotiations of previously held beliefs and ways of being in the world. In a context where the search for authenticity, or the 'search for God after God' has become increasingly important among both the religious and non-religious (Baker 2017a), spaces of postsecularity might offer new avenues for sacralisation. The translatability between 'God' and 'love', as discussed by Caputo (2001), as practiced in care for the marginalised might generate complex processes of re-sacralisation of compassionate in-commonness. Neoliberal governance mediates the expression of compassion through restrictive eligibility, deservingness, conditionality, sanctions, austere rationing and downgrading of support, breathing life into myths of individualisation, culpability and responsibility. This is not a celebratory call for charitable sentimentality but a political recognition of the potential, at least, in spaces of postsecularity for sacralisation of the 'other', whereby values of receptive generosity become foundational to understandings of responsibility. In doing so, we might better understand that the existential cultures of unbelief (Lee 2015) are not marked by disenchantment but entail a contested array of new forms of sacralisation – where embodied dispositions towards justice take on qualities of what is considered sacred. Last, the process of creating new myths that inspire solidarity, hope, and love are key aspects of performative postsecularity (Stacey 2017), and in doing so acknowledges the paucity of the myths we have lived by, become subject to, and have divided sensibilities of in-commonness into individualised silos of religion, identity, and belief.

Notes

1 This reading of curation comes from recent theological literature on inclusivity, creativity, and worship, particularly the work of Pierson (2012) and Baker J (2010). This theme of curation has also been recently developed in relation to new processes of partnership between faith groups, local authorities, and the public sector around themes of hope and spiritual capital in the work of Barber (2017).

2 We thank Wiley-Blackwell and Sage for granting us permission to re-use some material from two published papers: Williams A (2015) Postsecular geographies: theo-ethics, rapprochement and neoliberal governance in a faith-based drug programme. *Transactions of the Institute of British Geographers*, 40, 192–208. Williams A, Cloke

P, May J and Goodwin M (2016) Contested space: the contradictory political dynamics of food banking in the UK. *Environment and Planning A,* 48(11), 2291–2316.

3 Discussion is drawn from a two-month ethnographic placement working in a Salvation Army run 'Lifehouse' and drug programme. Daily involvement in the centre entailed working alongside staff and residents on the detox and rehabilitation wing of the building. Alongside participant observation, documentary analysis and extensive conversations recorded in a fieldwork journal, taped interviews were conducted with 14 of the centre's staff and volunteers and six residents at different stages of the treatment programme.

5 Political practices

5.1 Introduction

In this chapter we argue that postsecularity can be a useful ethos in the co-construction of political practices, identifying what we submit as potential progressivism within postsecularity, and paying particular attention to affect, tactics, and strategy. We focus on the ways in which the subjectivities and spaces associated with postsecularity connect with political scenarios, and offer an account that attunes thinking to the ethical and political potential of space-making in a way which problematises straightforward identification of assemblages as progressive or regressive. We illustrate how new forms of political protest can exhibit different forms of postsecularity, and how seemingly mundane participation in spaces of postsecularity can unleash capacities for, and tendencies towards, wider political mobilisation. Building on arguments developed in Chapter 3 and elsewhere (Cloke *et al.* 2016; Cloke *et al.* 2017), we argue that one of the key contributions of emergent postsecularity is to foreground debate on the politics of ethical subjectivity itself – how, why, and where are ethical sensibilities made, unmade, and remade; and how do these ethical subjectivities affect political mobilisations as they are simultaneously co-constructed by them.

Writing in a context of ecological threat, corporate dispossession, resurgent populism, and geopolitical and economic uncertainty (the Brexit and Trump phenomena come to mind for us Westerners), hopeful geographies might seem to be a scarce resource, or even politically naïve. Yet it is this context that brings into sharp relief the need for critical analyses of 'hope' and the formation of subjectivities oriented towards the Other. In recent work we have discussed how hope can act as a bulwark-against and opposition-to 'precorporation', the pre-emptive patterning of desires and hopes that eliminate ethical senses of value (Fisher 2009). There seems to be a convergence space within human geography around notions of hope in affective life, desire, and ethics, and the ways in which these connect to political-economies of neoliberal capitalism (Healy 2013; Jensen 2011; Vrasti 2009). The seemingly hegemonic grip of neoliberal reason (Anderson 2016) has produced 'compliant' subjects that acquiesce to the anxieties of neoliberal precarity (Firth 2016). What results is what some describe as a 'spiritual ennui' – a weariness towards social change and an emptiness of hope; a dissatisfaction, lack of interest, or existential boredom as the new normal in a world of phantasmagoria and simulation.

We want to suggest that this 'spiritual ennui' has damaging features on both the Left and Right of the political spectrum and that an ethos of postsecularity can encourage more progressive, ethical practices that create spaces and practices in common which transcend entrenched political alliances. At the heart of this torpor and division we are calling 'spiritual ennui' is a toxic propensity for dissociation reaped from what Coles (2001, 488) calls 'ressentiment'. On the Right there is a lack of willingness to engage in dialogue with symbolic Others because hopeful aspirations of unity in diversity are precluded by a lack of belief in possible change, or fear of the pain of change (Badiou 2003). This fosters an individualism marked by unbridled consumption and shirking of collective responsibility (Staeheli 2013). On the Left, there is a factionalist snafu brought on by a desire to simplify the enormity and urgency of the revolutionary task (Žižek 2011). This paralysis of reason on the Right and on the Left translates into recalcitrant edifices of political rhetoric that are violently protected, and then acted upon, for fear of doing nothing. However, these stances undermine the ability to share in the common experience of recognising the horror of the crises facing the planet. But how do we recognise these crises without fostering a defeated and fatalistic social dissolution that forecloses the subject's affective capacity to become enthusiastic about collective action (Žižek 2017)? Finding this commonality requires new practices of unlocking the ethico-political potential of the human vulnerability underneath unyielding rhetoric. For these practices of vulnerability to become possible, we argue that an ethos of postsecularity is necessary, and in this chapter, we will describe and analyse how a postsecular ethos can be generated in conjunction with political movements and how it can benefit them. How is hope in partnership and collective action generated and operationalised?

We are aware that this appeal to vulnerability can be problematic when trying to generate inclusive movements (Coles 2001). The presence of diversity assures the propinquity of power imbalances and appeals for marginalised people to be more vulnerable can often stick in the craw. Postsecularity is not about suspending the critical faculties which help people to speak unsettling truths to power. Rather it is an appeal for engagement that perseveres with the power relations in a diverse group in the hope that two things emerge: (i) a greater understanding of how to practically subvert oppressive power relations communally, and (ii) a sense of common purpose and solidarity.

We argue that exploring the construction of postsecularity is crucial for understanding and engaging with political mobilisations that bridge divides and exercise a vulnerability that enables the emergence of ethical investment across difference and a sense of common purpose. In this chapter we are interested in how the concept of postsecularity can illuminate the emergence of hopeful subjectivities that empower and are co-produced within political mobilisations. Although much scholarship on postsecularity and politics has focussed on how religion is being reasserted in the public sphere in novel ways (Cloke and

Beaumont 2013; Eder 2006), we are not concerned with examining the mechanisms of postsecularity because we think that religious actors are an *a priori* benefit to political life. Indeed, the religion/politics interface is a murky area in which political allegiances and religious meanings are negotiated and developed in an imbricated fashion, leading to both democratic, generous, and diverse polities as well as those marked by division and suspicion of symbolic others (Burbridge 2013; Connolly 2008; Hackworth 2010; Megoran 2013). Rather we are interested in analysing postsecularity because we have recognised elements and traces of it within recent political mobilisations and seek better understandings of how it is generated and what it makes possible. We would argue, however, that certain religious ontologies and practices do contribute to fostering postsecularity through their underscoring of transcendence (Cloke *et al.* 2016; McIntosh and Carmichael 2015). Although a connection with transcendence is not the preserve of religious individuals – a great number of secular philosophers, political theorists, and activists depend on notions of transcendence, even if it is an 'immanent' version of it (Badiou 2003; Critchley 2012; Levinas 1978; Sartre 2003; Žižek 2000) – in many of the examples that we draw on, religious actors have had a significant influence in foregrounding values or narratives rooted in transcendence, which have emerged as values that broad-based movements can muster themselves around. Therefore, in the rest of this chapter, part of the discussion will be comprised of an exploration of how religious ontologies and practices can make a positive contribution to progressive political movements.

However, first we want to discuss some examples of political collaboration that drew our attention to postsecularity; what might be called postsecular 'stirrings' (Cloke 2011). These mobilisations provide evidence of the coming together of religious and secular actors, demonstrating the co-production of mutual generosity and crossover narratives. One area of political activism in which we have recognised postsecularity and its effects has been in the resurgent hospitality and reconciliation movements in the USA that have opposed Donald Trump's divisive immigration policy (see Mindock 2017; Schaper 2016; Sherwood 2016). The resurgence of these movements that offer sanctuary to those the government would deport and condemn 'new forms of apartheid' (Žižek 2011, p. x) have demonstrated postsecularity in a number of ways. They have brought together communities of differing religions (and none) in order to provide sanctuary and protest injustice. They have demonstrated generosity by forming new communities of mutual aid that incorporate multiple (non)religious identities and motivations. This highlights a hope that despite difference in motivation and identity, community and conviviality can emerge that offer unconditional support to the vulnerable regardless of their identity and develop networks of activism – regarding mutual aid, protest, advocacy, and training – that function effectively despite motivational difference. We argue that this response stems from commonly held grief and rage that opposes powerful elites despite the 'uncompromising odds' (Megoran 2013, 145) of victory, and forms a commonly held commitment to sanctuary, a united and empathetic citizenry, and religious liberty (see Sojourners Website; Brueggemann 1978). Although these hospitality and reconciliation movements are not a new phenomena in the USA (see Day 1997; Phelps 2013; Van Steenwyk 2013), meaning that the current rejuvenation of the movement is a continuation of an extant tradition, the plurality and scope of the backlash against draconian immigration policy is unique to the moment and we argue that this uniqueness is partly generated by postsecularity. Although postsecularity indicates that people have the capacity to reach out to the Other in hope – rather than on the basis of an assured tactic – to develop unforeseen generosity and the ability to politically mobilise alongside difference, some might argue that this generosity existed already and that there is nothing uniquely postsecular about the current movement (Kong 2010; Ley 2011). However, as Megoran's (2010 and 2013) work illustrates, for significant political mobilisations – particularly of religious people – to occur, there often needs to be a reconsideration and shifting of religious values and narratives, a commitment to the Other despite the unpredictability of this engagement, and a shared grief or despondency that gives rise to a pluralistic search for new answers to a common political and affective plight (see also Cloke *et al.* 2012). This is why the focus of these revived hospitality and reconciliation movements has spurred opposition, not just to oppressive political regimes, but also to religious institutions that are dogmatic in their lock-step with the theological and political status quo (Rashkover 2017; Willis 2017). It is also why we recognise that, at this moment, these movements are being affected by a postsecular ethos that is engendering their collaborative practice, dovetailing with but also distinct from their previous activism.

A second political arena in which we have recognised postsecular stirrings has been in community organising, particularly in the UK. Organisations such as London Citizens have been heavily involved in Living Wage campaigns and efforts to increase social mobility (see Citizens UK 2018). Community organising initiatives like Citizens UK have brought together a great number of institutions with differing motivational bases, generating collaborations between faith groups, trade unions, and educational facilities. As Jamoul and Wills' (2008) work illustrates, similar work goes on in these organisations to develop a postsecular ethos of generosity and hope in the Other. They analyse how churches in London have shifted their practices in order to work out their faith through praxis, engaging in activism that seeks to protect vulnerable people in their locality, particularly through Living Wage campaigning. This demonstrates an ability to be vulnerable, exposing a cherished – and perhaps comfortingly settled – faith praxis to disquieting examination and the possibility of change. Again, shared frustration and despair regarding inequality leads to a collective cry against injustice and a shared outpouring of grief. In bearing with that pain together, new hopeful narratives based on common values emerge.

We have given these examples in order to highlight why we have been drawn to the concept of postsecularity and the unique analysis it can afford when studying political mobilisations. We have underlined what we think postsecularity is and what kind of political mobilisations it generates. In the rest of this chapter we will demonstrate via different case studies how the three modalities of postsecularity outlined in Chapter 1 can be facilitated, outlining how each case study generates a different postsecular modality with particular political utility. First, we will discuss spaces that engage marginalised groups in participatory arts practices. This analysis will illustrate how co-productive spaces which focus on enjoyment and interaction can engender generosity across difference by facilitating an ethical proximity that highlights the common and profound human experiences expressed through participatory arts practice (Olson 2005). These spaces are of political consequence because they consolidate a shared affective experience of injustice and foster a sense of hope in collective action through unleashed creative power (Diprose 2015). We tackle this modality first because we argue that this sense of in-commonness is core to the functioning of the other modalities of post-secularity. Although political demonstrations of pre-formative ethics can easily falter, they can also create an affective capacity for forbearance and ethics that the political negotiation of receptive generosity and rapprochement rest on. The post-secular modalities of receptive generosity and rapprochement cannot exist without a pre-existing hope that engaging the Other – that which is not understood or is found threatening – can yield a benefit. This hope can also be generated by Other-embracing ontologies (see Augé 1998; Badiou 1997; Blok and Farias 2016; Brewin 2010; Eagleton 2009).

Second, we will explore how new forms of religious community encourage techniques-of-self that can increase and extend receptive generosity, augmenting postsecular ethical capacity and generating connections between more mundane spaces of dialogue and conviviality, and arenas of political action which blur definitions of secular and religious space. Third, we will analyse how the Occupy movement in London and New York generated new forms of protest and narratives of protest, and how the postsecular modality of rapprochement Occupy exhibited can engender broad-based political mobilisations characterised by increased commitment to build coalitions across difference and allowing new common goals to emerge rather than be ideologically proscribed. To conclude the chapter, we will suggest new arenas of research and activism that postsecularity can enrich, along with alternative ontological frames that can enhance the understanding and praxis of postsecularity.

5.2 Developing generosity: postsecularity and prepolitical 'bubbling-up'

In addition to the 'stirrings' already illustrated, we argue that postsecularity can be detected in a wide range of political movements with peace activism (Epstein 2002; Tosi and Vitale 2009), the Umbrella Movement in Hong Kong (Chan 2017; Tse 2015), and the Arab Spring (Barbato 2012; Mavelli 2012) being good examples. However, such political movements can be guilty of stifling dialogue and change, for fear of not having defined goals, strong leadership, or routes to conventional offices of power (Juris 2005; Sullivan 2005). As Button (2005), Coles (2001), and White (2016) argue, inability to exhibit generosity, vulnerability, and hope often leads to movements excluding the very people they are supposed to help, foreclosing the possibility of broader, more powerful coalitions, and diminishing the organisational and rhetorical agility necessary to see political goals to completion. The ideological purity that is protected by the lack of a postsecular ethos can also result in missing opportune political moments because there is no openness to moving towards consensus.

Although we argue that postsecularity is politically useful, we do not want to sublimate a postsecular ethos here. Political movements are marked by contingency. Leadership and hard boundaries can sometimes make a space more accessible for vulnerable people by excluding actors who are suspected of having malevolent agendas. Excluding harmful agents illustrates that the extension of generosity must sometimes be limited rather than being unconditional (Routledge 2003). Additionally, uncompromising commitment to political, ethical, or spiritual idea(l)s can be a quality that helps a political movement break through to a broader consciousness, grow in its capacity, and begin to affect society in the way it has set out to do (Alinsky 1971; Chesters and Welsh 2006). However, we present postsecularity here as an ethical approach to politics that does not feebly submit to overbearing or oppressive power structures in an effort to be generous in a servile way or to be naively vulnerable. We posit it as an approach that sees the possibility and life that can come from generosity and vulnerability and is willing to consider hopeful risks when politically engaged, recognising the chaotic openness of reality-as-it-is, rather than defaulting to defensiveness, ressentiment, and purported certainties (Holloway 2011b). Postsecularity is an ethos that makes a range of hopeful tactics and imaginations possible within the political realm, adding to extant political techniques whilst simultaneously suggesting a pragmatically fluid approach to activism that aims for accommodation of difference within a movement and the emergence of a collective sense of purpose (Holloway 2013; see also White 2016).

But how does this generosity and sense of purpose emerge or bubble-up? As outlined in Chapter 3, postsecularity has been identified as a prepolitical emergence of feeling in-common. In the case of foodbanks (Cloke *et al.* 2017; Williams *et al.* 2016), this emergence is defined by whatever affective state a volunteer, foodbank employee, or client comes to the foodbank with being added-to, superseded, or transformed by a collective sense that the hunger and food insecurity that clients are enduring is wrong and unjust and that this must be made clear to a wider public and to governing institutions. The rage and grief that energise this declaration generates unexpected unity between people who have come to the foodbank with very different motivations, affective states, and worldviews. This affective rail against injustice binds and transforms the whole gamut of actors co-constructing the space of the foodbank; clients who have been made to feel deep shame by the media and foodbank employees, conservative volunteers who have at one point felt that foodbanks should be the norm in food politics, and more radical employees who have a problem with the politics of foodbanking but feel compelled to do something to help those in need. Cloke *et al.* (2017) have argued that it is in the mundane spaces of the foodbank, in the 'doing something about [something]' (Cloke 2002, 589), in engaging on a regular basis and forming relationships with people who are suffering injustice, that generosity and vulnerability emerge in ways that can transform political recalcitrance into feeling-in-common. Through encounters with the Other, Levinas argues, '[t]he subject, whilst preserving itself, has the possibility of not returning to itself' (1978, 165). It is this engagement with the Other and accompanying postsecular affective transformation and bubbling-up that we are interested in rather than the promotion of an affective openness to transformative encounters that reject any form of commitment to the wellbeing and valid experience of Others and in so doing become committed only to novel or pleasurable affective experience and unrestrained relational experimentation (Coles 2001; Caygill 2002).

However, as Sullivan (2005) and Juris (2005) argue, the kinds of spaces that are responsible for the emergent generosity, vulnerability, and hope often needed to generate new ways of being together as well as feeling and purpose in-common are spaces of enjoyment, play, conviviality, and experimentation (when they are thoroughly ethically safeguarded). Their separate papers on the World Social Forum both highlight how it is often the spaces around the edges – parties, cultural displays, shared meals – of the 'real work' of the Forums' deliberations that generate the greatest sense of hope, empowerment, transformation, and new senses of togetherness and purpose. We are interested in these mundane spaces of informal connections and enjoyment as they seem to be a key source of generating hope that engaging the Other can yield dividends. This hope in the Other is the basis of a postsecular generosity and vulnerability and allows for the bubbling-up of common political purpose. Although these spaces may seem mundane or politically insignificant, they can in fact create a foundation upon which political mobilisation can be built.

This evidence of the potential generosity arising from postsecularity is reinforced in Cloke and Sutherland's (forthcoming) research on participatory arts practices with homeless people. These kinds of practices with marginalised groups could be argued to be 'missing the point'. Many writers who oppose the revanchist remoulding of the city that has excluded various marginalised groups – and particularly homeless people – argue that the only way in which solidarity can be extended to marginalised people is by holding the state to account for its abandonment of them (Davis 1990; Mitchell 1997). However, Duff (2017) and Smith (1992) argue that arts practice can help homeless people to reappropriate their bodies from oppressive systems, maximising their autonomy and ability to make choices that construct meaningful lives. It can increase access to networks of care and welfare, to practices of self-actualisation that help them to survive, resist, and subvert revanchism, and expose broader publics to the voices of homeless people and their allies.

However, we want to identify participatory arts practice as holding postsecular potential; the inclusive engagement therein can be extremely useful for transforming relationships within a group and enabling new affective topologies to emerge which can guide a group towards collective action (Diprose 2015; Douch 2005; Powell 2004). As Richardson (2013) highlights, spaces where people generate artistic pieces and performances collectively engender relational instability, making space for participants to experiment with and experience new identities and to reconfigure self-to-self and self-to-other linkages (see also Kershaw 1992). This can be useful in generating the generosity, vulnerability, and hope in the Other needed for postsecular bubbling-up to occur. However, it must be noted that there is a topology of power relations to negotiate within these performance spaces. Ethical assessment of these spaces should recognise that there are degrees to which people should be expected to shoulder the emotional labour of generosity. Notwithstanding, Askins and Pain (2011) emphasise that particularly when engaging marginalised groups, the non-verbal aspects of participatory art are vital tools for generating a sense of inclusion, participation, and self-worth through processes that allow people to be vulnerable and trusting. Vulnerability and connectedness are facilitated at an angle through art, rather than directly through dialogue, which can sometimes fail in its ability to help people to express complex amalgams of conflicting emotions to one another (Ruud 2004). They assess what becomes possible when a group of diverse kids comes together to create a participatory art space in what is usually an ethnically segregated area. Although the outcomes were modest in their scale, longevity, and ability to be measured, by letting the kids interact with one another in designing the aims and processes that the art space would be geared towards achieving, moderate success was realised in the forming of friendships between kids from previously segregated backgrounds. By having fun together in making artwork, the pleasure of one another's company was able to transform suspicion of difference into a sense of possibility regarding the co-production of mutual enjoyment. In addition, further points of resonance were generated by helping the kids to identify common interests and reasons to re-evaluate fears around difference. Discussions were had about football players the kids liked, food they enjoyed, and things they liked about their locality which all underscored the positives of intercultural exchange. Senses of difference and sameness were re-negotiated between the kids through the artistic process, and a new sense of generosity could emerge between some of them, leading to greater intercultural integration. This emergence of generosity and warmth towards someone who is previously perceived as the Other is the transformation at the core of postsecularity and creates the conditions for the bubbling-up of things in-common. For these kids it was a new sense of the enjoyment and enrichment of life that can stem from intercultural relationships, generating a new way for the group to be together and improve their ethics. This is important political work, changing relationships at the cultural and social level. However, artistic spaces can also give rise to political mobilisations that are less about a new sense of how to be together, and more about how society more broadly needs to be altered.

For example, Kumm's (2013) research focusses on how artistic practice can give the subject the ability to process trauma and forge a new identity that is not bound to their victimisation. However, as Boeskov (2017) and Diprose (2015) point out, it can also help people to interrogate and critique ways in which they have been politically oppressed. For example, Finley and Finley's (1999) study of the poetry scene in New Orleans highlights how collaborative spaces that encourage the engagement of homeless people, can allow a collective sense to emerge both of the validity of homeless people as artistic equals and of fury at the reductive and marginalising way in which they are governed. By legitimating homeless people as artistic equals in the accessible arena of the streets, it makes it possible for a broader public to treat them as valuable contributors to a public activity whilst simultaneously giving them the option to develop social and artistic connections that they can use to access spaces such as clubs, bars, and galleries that revanchist policy strives to exclude them from. The scene is formed by ethical and artistic connections within and beyond the homeless community, co-creating new narratives about homelessness that represent homeless people as human and complex, altering fellow artists' and audiences' perceptions of homelessness and chipping away at revanchist rhetoric (Finley 2000).

Identification of the postsecular potentiality of participatory arts practices should not, of course, ignore the political and ethical quandaries that arise when trying to facilitate these spaces. Marginalised people can be patronised by attempts to facilitate practise they are already engaged in (Kidd 2009), institutional patronage of such schemes can wrest interpretive control of the art away from marginalised artists (Kershaw 1992), and in bringing diverse groups together, conflict rather than integration can sometimes be the outcome (Boeskov 2017; Askins and Pain 2011). However, the potential that these spaces have to co-produce the ethical investment across difference and the bubbling-up of feeling in-common which characterise postsecularity means that they have to be regarded as significant spaces capable of shifting broader social, cultural, and political landscapes. What is more, as Kumm (2013) points out, arts practices often help people to access the parts of themselves that are mysterious, transcendent, or spiritual; the deep desires that are ineffably greater than the sum of discernible psychic parts. The performativity of art spaces conjures up the mystery and ethical sensitivity of religiosity, generating the possibility for it to come into a position of postsecular negotiation with the structure and political reasoning of secularism.

These spaces of collaborative arts practice make hopefulness in engaging the Other possible. By engaging in collaboration, creativity, and enjoyment, participants build trust, ethical and emotional proximity to one another, and co-create new affective topologies that resculpt intercultural relations, generate new senses of common purpose, and posit alternative collective futures (Yates and Silverman 2016; Holloway 2013). However, the hoped-for outputs of these spaces – particularly greater intercultural appreciation within the group and the bubbling-up of a common affect with a political target – are not guaranteed. The collaborative practices create space for their possible emergence through inclusion and engagement but do not underwrite their materialisation (Askins and Pain 2011; Davis 1997). In response to this uncertainty, some political, artistic, and religious practitioners are being more direct in their encouragement of ethical investment across difference within diverse groups through praxis that responds to 'speculative ontologies' (Moody 2012, 183).

Speculative ontology emphasises the absence or opacity of the Big Other or Master Signifier (Bryant *et al.* 2011; Žižek 2000) and therefore the meaningless-ness of discerning a totalising meaning to life or telos (see Caputo 2001; Deleuze and Guattari 1994; Rollins 2006; Taylor 1987; Žižek 2000). Rather than leading to lassitude and ennui, these thinkers have embraced the difficulty of construct-ing a stable meaning to life, positing it as a turbidity from which modes of being that engage hopefully with the Other can emerge. As Žižek (2000) puts it, the competing demands that society makes upon its governing structures or status quo, and the grief that emerges from the everyday violence of – for instance – the currently predominant model of liberal capitalist parliamentarianism, render the social order absurd and obsolete (see also Brueggemann 1978). Upon accept-ing the reality of the obsolete social order, the subject has no option but to forge a new sociality with those who have similarly felt aggrieved by the futility and oppression of that 'order' (see Badiou 1997). Žižek (2000) argues that this grief in-common is analogous to the Holy Spirit in Christianity. The early Christians accepted the futility of the previous religious and socio-political order. In griev-ing together at the oppression wrought by the current social order – whether legalistic Judaism, Pax Romana, or capitalism – this community of social out-casts commits to one another's wellbeing by refusing the oppressive relations of the social order and committing to enacting and disseminating a speculative ontology. This speculative ontology prompts a hope (or – as Žižek (2000) posits – Holy Spirit) that is enchanted by an alternative vision of the future. The Holy Spirit refuses the current social order and despite the fact that it presents no con-crete solutions as to how a new social order will emerge or what it will look like, it insists that dedication to the community of the outcast and aggrieved is the only thing that can lead to the emergence of a social order in which ethical investment across difference and the bubbling-up a sense of common purpose is possible. Žižek (2000 and 2017) argues that in a shared experience of powerless-ness – when the hegemonic social order has been exposed as fraudulent and there is no 'plan B' – there is a phenomenologically identifiable shift within a group, where an affective topology marked by hope, humour, love, and solid-arity bubbles-up and points the way forward despite very little rational grasp of what to do next.

However, because there is an absence of a Master Signifier or telos, maintain-ing the possibilities that emerge from embodying speculative ontologies must be marked by an iterative praxis of iconoclasm – so as not to generate an inviolable hierarchy – and openness to the event (Badiou 1997). The event in these specu-lative ontologies is an experience that irrupts into everyday life, altering the sub-ject's experience of it fundamentally and demanding that life be led in such a way that reflects to those who have not experienced the event the new possibil-ities it suggests (Badiou 1997). Žižek's ideas about speculative emergence of hope and solidarity through shared political indigence represents one such framing of the event, which breaks in and unsettles the current order. The notion of fidelity to the event turns up in a range of both religious and secular speculative ontologies. A religious example is presented by Taylor (1987), where fidelity to the event represents itself as a/theology; a recognition that imitating or experiencing the divine through dedication to discursive prescriptions is unreliable and that it is in fact experienced serendipitously in ways that upend attempts to categorise it. A secular example can be found in Deleuze and Guattari's (1994) discussion of combining immanent factors that produce something experienced as novel and often destabilising or contradictory to the social order.[1] Although there are different notions as to what constitutes an event, the ontological basis of an event, and how it should be responded to, it represents a key concept in the structuring of a range of speculative ontologies. Speculative ontologies follow the pattern of accepting that the current social order is defunct and awaiting the emergence of an event that shifts the experience of and framing of reality and points the way to alternative futures.

Holloway (2013) has argued that the overlap between these kinds of ontologies underscores the possibility of progressive futures. The parallels between speculative ontologies across secular and religious thought both structurally foreshadow and point to the possibility of postsecularity. Indeed, postsecularity and speculative ontology share an idea about the structure of political and subjective transformations; both put stock into the embrace of uncertainty around meaning, followed by openness to the incursion of an affect of unexpected possibility. These parallels prompt questions about who can be brought together in a shared experience of a crumbling social or signifying order. If religious and secular thinkers – with varied motivations, intellectual projects, political ideologies, and visions of the future – are conversing about similar patterns of generating subjective and political progress (see Davis *et al.* 2011; Hauerwas and Coles 2008; Depoortere 2008; Rollins 2006), what might this conversation look like between political activists? How might a shared assessment of a political quandary as evading meaning or foreseeable resolution lead to a common space in which to experience ethical proximity through mutual recognition of reality as manifold and the bubbling-up of an affective topology of hopefulness?

To give an example of how to create this kind of common space, research by Moody (2010 and 2012), Ganiel (2006), Ganiel and Marti (2014) highlight how speculative ontologies have been introduced into the praxis of religious performance artists, generating a postsecular emergence of ethical proximity and a shared sense of purpose by exposing the internal contradictions or hierarchical modes of theological thought and organised religion. They both focus on the Ikon community in Northern Ireland, and the speculative ontology of a/theism that the collective tries to enact. Ikon's commitment to a/theism emphasises that in practicing and believing in Christianity, God is experienced as much as an absence as a presence, and that God is beyond categorisation. Christianity is not about obeying religious precepts but living as if God does not exist; or rather that a stable version of God does not exist, and this generates the possibility of living in a way that both evades psychological suppression and is ethically aware of the co-constituting relationship between self and Other. In this way, Ikon curates performance art that destabilises people's notion of self/Other boundaries and of God, engendering a postsecular ethical proximity through collective acceptance of disavowed social and religious authority, creating space for new experiences and meanings of God and the religious collective to emerge. For instance, one of their performances involved an actor reading out what one of Moody's (2012) interviewees describes as a 'gorgeous' (p. 194) sermon. Then, the original recording of the sermon was played as delivered by the politician and Presbyterian minister, Ian Paisley, troubling listeners by attaching an ominous cadre of political resonances to the words. The discomfort that this performance engendered encouraged the audience and performers to question the cultural co-ordinates they use to locate truth and beauty. This generates postsecular questions about the deeper desires and fears of politically problematic Others. What happens when we identify with them? Can we be generous to objectionable characters by recognising their pain and longing, generating points of dialogue on this basis so as to query their fundamentalism and ressentiment? How can we practice generosity with these Others without supporting their harmful political agendas? These questions are pertinent to postsecularity, creating the possibility of ethical proximity across difference, and making space for new political imaginations and tactics to emerge.

In Moody (2012) and Ganiel's (2006) narrative of Ikon, we see the postsecular potentiality of art practice coming to fruition and pushing through from an open-ended space of generosity to one which actualises that generosity by encouraging participants to think about how the new ethical proximity they have with the Other might enable more widely-felt political imaginations and mobilisations to emerge. Although these practices take place within a religious setting, they illustrate how there can be concrete practices which could be carried across into political and ethical spaces in order to encourage postsecular generosity across difference and make space for the bubbling-up of new political possibilities. When there is a common experience of enacting speculative ontology, various notions of sameness/difference are drawn together and renegotiated. This is a particularly potent example of postsecularity resculpting the self-to-self relation of religious identity and transcendence and this having knock-on effects on the self-to-other relation of pluralised and secular political engagement. Although these spaces do not rely on their participants or facilitators to have an understanding of speculative ontology, Holloway (2013) ponders how reworked practices of ritual, art, and collaboration can be operationalised to the appreciation of various champions of speculative ontology in order to generate further postsecular political mobilisations. This, in turn, generates further reflection upon how best to imbue different spaces or political scenarios with different speculative ontologies. Why might differences between Žižek's (2000) speculative ontology and Taylor's (1987) be important? And may different speculative ontologies be better applied in different situations; for instance, in a homeless shelter, or a religious gathering, or a political debate? What practices do different speculative ontologies make possible?

Questions about the content and practical implementation of speculative ontology in the service of postsecular ends are crucial because speculative ontology is equally capable of ushering in the dark side of postsecularism. In much writing on speculative ontology, the upwelling of hopeful affect – the in-breaking of the event – in a collective, emerging from their new-found ethical proximity is often ill-defined. Although thinkers like Deleuze and Guattari (1994) and Nancy (1994) stress the importance of being open to the event as the 'raw opening of possibility' (Coles 2001, 497), Coles (2001) underscores that this lack of definition can lead to a problematic refusal of power relations and positionality.[2] Coles (2001) uses Anzaldúa's (2012) work to stress that generosity can be extended, even to seemingly problematic Others – the subject can be open to an experience of recognising the humanity in and finding ways to work with even the most evil person – but that this must be mediated by a polyvocal assessment of the ramifications of this extension of generosity. Anzaldúa (2012) endorses the Navaho saying '[o]ut of poverty, poetry; out of suffering, song' (p. 69), stressing that the emergence of new political possibilities depends on taking risks on people, being open to hope rather than suspicion through recognition of common disappointment and pain. Sometimes – Anzaldúa argues – these risks can seem like bad choices, moving into uncharted, unstable territory but nonetheless creating progressive opportunities for marginalised people. However, they can also include great danger, making vulnerable groups yet more vulnerable to exploitation and harm by hegemonic power. It is therefore crucial to scrutinise the event – and its attendant imaginations regarding praxis – for its political and ethical consequences. Caputo (2001), Rollins (2006), and Žižek (2000) argue that the consequences of the event should be greater love for the Other, but love is a complex concept that they show little inclination to define, and – as Anzaldúa points out – to develop a progressive practice we need to be crossed by many Others and be attentive to the topology of power between ourselves and this multiplicity. McIntosh and Carmichael (2015) argue that this openness to the Other on its sliding scale of ethico-political problematics, and the reality of our own transformation by the novelty of the event requires something beyond ethics or affective openness. They contend for what they call 'spiritual activism' (p. 11), a transpersonal psychology of reflection upon the tense relations of reality-as-it-is, negotiating between an accepting relation of self-to-self which refuses to suppress the personal experience of the event (and its concomitant sense of hope, possibility, and excitement), and a care for multiple Others (see Augé 1998; Žižek 2004). Spiritual activism represents a divining rod for the operationalisation of speculative ontology in postsecular practice with its risky engagement of diverse Others, and its relation to the unknown possibilities of emergent affect.

However, creating space for ethical investment and the emergence of politically directed affect, sometimes is not enough to generate the postsecular political outcomes that we are interested in in this chapter. As Askins and Pain (2011) point out, these participatory spaces can engender greater warmth and vulnerability between participants but fail to deconstruct political ressentiment. As they say (following Valentine 2008), 'courteous encounters do not automatically translate into respect for difference or alter dominant social values' (p. 805). We need to be up front about how these kinds of spaces, which rely so much on the irruption of an event, can be sites of ethical and political disappointment (Braun 1997). However, as our example of the Ikon collective shows, sensitive, and rigorous facilitation can encourage the becoming of political and ethical subjectivities in these spaces along a predetermined progressive trajectory (see Boeskov 2017). In the next two sections we want to focus on the two other modes of postsecularity that we outlined in Chapter 1 and are threaded through the narrative of this book. We argue that receptive generosity and rapprochement are concepts that delineate the development of political responses established upon the ethical investment and affective emergence in-common of postsecularity. In the following two sections we illustrate how these concepts can help political practitioners to facilitate spaces so as to provide them with an additional set of tools for converting vulnerability to the Other and affect in-common into concrete political action. We begin with receptive generosity.

5.3 Receptive generosity

As outlined in Chapter 1, receptive generosity is a way of practically grounding the ethical proximity and affect-in-common of postsecularity in political scenarios. The concept of receptive generosity highlights the iterative process of subjects listening to one another's differences and being open to relational transformation by reflection upon the legitimacy of the friction that these differences engender in the subject's self-to-self and self-to-other relationships. Following Coles (1997), if a group commits to privileging no single ideology over another, this can facilitate a praxis of enhanced dialogue across difference through reflection upon shared experiences and values that may be useful in the political moment despite their lying outside of the typical modus operandi. This not only enables timely tactical pathways but underscores the possibility of mutuality in circumstances where it is unforeseeable. Experiencing this possibility, despite seemingly unfavourable circumstances, can encourage subjects to experiment further with reflection upon self-to-self and self-to-other relationships, generating a greater sense of hopefulness and an expanding, fluid slate of political tactics and rhetoric. For Coles (1997), receptive generosity augments the possibility of constructing broad-based movements marked by participative and agonistic democracy whilst establishing links across political divisions.

Significant socio-political theorists such as Connolly (1999) and Foucault (2005) have devoted substantial attention to understanding the concrete practices that enable the reflexive transformation inherent in receptive generosity, and both refer to key examples from religious and spiritual traditions. Foucault in particular is interested in the spiritual practices espoused by Greek philosophers in the pursuit of truth (such as rites of purification, concentrating the soul, withdrawal, and endurance) and in the techniques of self (like prayer, contemplation, and confession) used by Christian monks to evaluate themselves morally and alter their feelings and practices. However, more so than Foucault, Connolly (1999) is interested in how these practices have political utility. Developing a progressive reading of Nietzsche (1974), Connolly argues that all political and moral positions – both religious and irreligious – bring metaphysical baggage to bear upon their predilections, formed by the inscrutable intersubjective production of confluent affective and discursive resonances. Connolly argues that Nietzsche's thought harmonises with an a/theistic ontology (1999, 54), recognising the opacity of positionality, and the impossibility of fully understanding the fragmented and folded self (see also Rose 1997) which is animated by '"concealed gardens and plantings" below the threshold of reflected surveillance' (Connolly 1999, 28; quoting Nietzsche 1974, 9). Subjectivities and ethico-political positionalities are constantly becoming and co-produced through an immense and largely spectral network of power-relations, rendering any claim to metaphysical primacy unreasonable. A subject's becoming is dependent on sources of knowledge and structures of power that they might deem irrelevant or pernicious but that generate the material or discursive conditions upon which the subject is able to base their ethico-political position. Given that all subjects are riven with blind-spots and contradictions, Connolly suggests that ethical engagement can bridge political tensions when sensitive dialogue focusses on the variegated quality of suffering that pervades diverse groups. Dialogue about the different qualities of suffering can augment a sense for what a group has in common – for example, oppression by hegemonic elites – whilst also subverting egotistical ressentiment, cutting through problematic positionalities which obfuscate pain, fear, and vulnerability. However, attaining this level of mutual vulnerability which – as Coles (1997) suggests – emerges from critical reflection upon ideological priorities, requires the individual and collectively performed psycho-spiritual techniques of self that intrigue Foucault (such as contemplation and corollary practices). In order to embody a receptive generosity that increases openness to relational transformation, self-reflexive techniques are needed that enable the subject to wrangle with their deepest fears and prejudices and enhance their affective capacity to engage with difference. These techniques can be responsible for both the initiation and extension of the postsecular indices of ethical proximity and emergent affective solidarity.

In this section, we are interested in exploring how techniques of self can implement receptive generosity in order to launch emergent senses of in-commonness into a praxis of broader political engagement. We will be using examples from Sutherland's (2016) research on contemporary Christian communities that embody a postchristendom attitude[3] (see Bartley 2006; Van Steenwyk 2013), breaking from institutional patterns of religious expression in favour of forming relationships that: (i) create space to de/re-construct belief and faith, and (ii) imagine and support faith praxes with political implications. Often, religious communities have been evaluated as on-the-ground expressions of top-down religious authority (Dittmer 2008; Hackworth 2010; Yeoh 1996), or as sites where multiple actors contend with one another to inscribe a space with their preferred religious meanings (MacDonald 2002; Southern 2011; Vincett 2013). These are not redundant epistemological approaches, but they only present part of the picture. What about when religious community formation has less of a focus on implementing diktats, and more of a focus on facilitating spaces of dialogue that enable subjects to explore their religious formation and that of their peers (see Conradson 2013; Heelas and Woodhead 2005)? What about when religious community formation has less of a focus on inscribing a particular place with religious meaning, and more of a focus on preparing subjects to notice religious meaning emerging serendipitously across multiple, sometimes surprising, locations (see Holloway 2003; Lane 2002)? Sutherland's work draws parallels between religious and political communities, highlighting a series of debates in political geography about how activist communities negotiate the melding of inclusion and democracy with direct action and broader solidarities (Blühdorn 2006; Chesters and Welsh 2006; Della Porta 2005; Häkli and Kallio 2014; McDonald 2002; Pickerill and Chatterton 2006; Poletta and Jasper 2001; Routledge 2003). Similar to political communities, although religious communities are suffused with contestation and problematics, they can also induce the more generative and hopeful processes that political geographers have been expounding for many years. In order to help their members feel safe to explore their religious formation, they try to foster inclusive atmospheres and democratic engagement. Aspirational inclusivity is the ethically attuned environment in which discussions can be had about diffuse religious geographies – drawing on sources of religious significance that also have political implications, like direct action (Reuver 1988; Roberts 2005), hospitality (Claiborne 2006; Day 1997), and protest (Bloomquist 2012; Howson 2011) – as well as enabling the community to reflect upon its *raison d'être*; is it more concerned with facilitating communal political praxes or empowering individual ones? In order to flag the progressive potential of these communities, Sutherland highlights the ambivalent processes inherent in religious subjectivation[4] and underscores the helpfulness of entwining a postsecular ethos of receptive generosity and its attendant techniques of self into processes of religious subjectivation so as to enhance the capacity of communities for democratic inclusion and political motivation. By analysing Sutherland's autoethnographic work with some of these communities, we will demonstrate how receptive generosity functions so as to facilitate postsecular ethics and emerging affect in-common towards political activism.

The first example we want to draw on is from Sutherland's experiences at ecumenical youth conferences that were organised by Christian social justice organisations. What drew Sutherland to these conferences was the temporary community they constructed, centred around a common rejection of hierarchical theologies. Whilst not necessarily opposed to universal truths, these conferences were bound by a sense that religious institutions can be needlessly exclusionary in their policing of purportedly right and wrong theology and morality.[5] The *raison d'être* of the conferences was to create spaces where subjective experience of the divine and the political hermeneutics of liberation theologies could be brought to bear on theological discussion as well as scripture and tradition. The spaces they created were supposed to generate dialogue, personal reflection, and pathways for experimenting with religio-political praxis. They already exhibited significant postsecular traits; people were coming together – bound by a rejection common of religious authority – to do the ethical work of transpersonal reflection across different subjectivities and praxes. Additionally, through their foregrounding of politics and praxis (workshops, talks, and rituals featured issues such as direct action, nonviolent resistance, ethical consumption, and protest), there was an attempt to engender an expectation of emergent political affect and action.

Below is Sutherland's account from one of the conferences he attended on the effect that techniques-of-self had on his ability to connect with a broader pallet of political practices.

Box 5.1 Autoethnographic reflection on receptive generosity

When I first arrived at the conference, I was excited to engage with people who had been struggling with some of the same issues as I had been. For a long time, I had been disappointed with the level of engagement with social justice in churches that I had been a part of, as well as the inflexibility of their theology. This sense of excitement was heightened when I first entered the main hall that the conference was taking place in and was met with a plethora of signifiers of activism and ecumenism mixed together in the space from the outset. There were tables littered with white poppies and books about nonviolent resistance, a craft corner with provocative quotes on the wall; a mixture of bible verses and facts about the arms trade, and right across the back of the hall, a massive blanket made up of squares covered in messages promoting peace. There were a couple of Franciscan monks chatting and strolling about, some women from a Christian communist movement doing some artwork in the corner, and a band rehearsing around a piano.

What I found most interesting about this space was the way that the song the band were rehearsing began to stand out and shape how I was feeling. It was a song I knew, one that had frequently been played in some of the churches I'd gone to in Glasgow. The same ones that I'd struggled with in the past. But instead of instilling me with a sense of trepidation, the song acted to make me feel at home. I felt at ease, not so manically energised to try and engage with all the activists that the hall was slowly filling up with. I noted this down because I thought it was strange but didn't know what to make of it. It was a while before things got kicked off properly, so I left the hall to stretch my legs.

When I returned, the conference had officially begun, and after a brief introduction, the first collective activity was singing together. Again, the songs were ones I knew but instead of feeling at ease, I felt like I couldn't participate fully. The atmosphere was distinctly flat. People weren't singing with much enthusiasm and this continued for the extent of the session. This left me feeling disconnected and out of place. The excitement I felt when I first entered the venue had dissipated as the solidarities that I had expected to feel instantaneously seemed to grow distant to me.

Whilst reflecting later, it struck me, as I was thinking about the way I had reacted to this worship experience, that even though I was not consciously doing this at the time, I'd acted as if I was disappointed because not everyone was performing the worship space as I expected them to. I reflected that this reaction was not particularly in line with what my theoretical theology of worship was. Did I think that everyone, from every denomination should respond in the same way to a practice that is relatively confined to a particular brand of Protestantism? Of course not. Their minds and bodies aren't trained to be receptive to the ecstatic in the same way as mine's is.

However, in hearing a familiar song when I first entered the conference space – I was primed with certain expectancies of what it would engender in the group experience. I felt comfortable because I presumed that something familiar was being prepared. As I mentioned, I had struggled with the churches I had previously been in that used this music. However, by using this music these churches had – despite my misgivings – created affective environments that I found enjoyable. They'd enabled me to practice what Julian Holloway (2003; following Massumi 1997) calls collective individuation – where the surrounding environment is enlisted by a religious individual in order to craft their own experience of transcendence. Although struggling with these churches, they had also been places where – through the music – I'd felt my own connection to something in excess of immanence. I'd argue that because of these experiences I was still holding on to a half-suppressed thought that even though I had a lot of problems with these churches, they must have been doing something right to enable me to experience these numinous feelings and that everyone should fall into line with that.

But it's not that the spaces where I'd had these experiences before had been doing something right. They'd taught me to become open to something excessive in a particular way, and in a particular environment, one in which it seemed that whatever affect was being crafted by the music was having a similar affect on all participants. In fact, the affect which – although experienced personally – was supposedly happening to the congregation, was in fact equally created by the congregation through their enthusiastic participation in sung performance.

However, these reflections came to me later, and I was prompted into reflecting because of my continuing experience in the same worship space. At the end of the singing, those guiding the session from the front encouraged everyone to 'share the peace'. Sharing the peace usually involves going around the room shaking hands with or hugging people and saying 'peace be with you' to them. This is supposed to incorporate people being reconciled to one another, working through conflicts or forgiving one another, being an outworking of active pacifism; putting relationships right rather than settling for the absence of conflict. In reality, people rarely take the time to go further than a handshake or hug during a service.

When it was announced that we should share the peace after the worship session, this was the last thing that I wanted to do. I was frustrated with myself and the environment because the affinities that I had expected to feel instantaneously did not emerge. When people started to share the peace, I stood up and shook hands with the people immediately around me. This is partly because this is how I had experienced this ritual before (I had never seen people travel far from their seat) but also because I had little desire to engage; I wanted to stew in my discontent. However, when I looked around it seemed as if everyone was trying to hug everyone else in a room of approximately a hundred and fifty people. Observing the excessiveness and joyfulness of this collective act began to erode the annoyance that I was clinging to. It struck me as a powerful theological symbol; it was a visible manifestation of the desire to perceive everything as interconnected, a recognition of the networked nature of the subject and of a desire to be attentive and caring towards the relationships which both wittingly and unwittingly sustain it. Although I could feel my emotions slowly shifting in response to this, there was a more sudden shift in thought and affect within this more gradual drift. Layered within my response to the ritual was a sudden thought – that came with clarity and a sense of stillness – that, more than just being a pleasing symbol, this ritual was facilitating an authentically spiritual moment. This in-breaking thought seemed to me a kind of transcendence. I am personally hesitant about connecting ritual with transcendence, because the affect that rituals can conjure is a mysterious excess that can so readily emerge from the blending of completely immanent actors (the effect of music on a worship space is a prime example). However, when I experience moments of such sudden alteration in my affective and cognitive stream – although I hold off from making quick judgements about them – they usually prompt me to reflect upon what they might say about my frame of transcendence and what that means for my future praxis.

I had not shifted completely from my anger and frustration, but this moment of religious significance was a key factor in shifting me towards reflection. When reflecting, as well as recognising that there were contradictions in what I thought and felt about the space, this transcendent moment was crucial in highlighting to me that despite my own sense of distance from what was going on, I might be missing out on something worthwhile. It also highlighted to me that even if the collective was practicing in a way that I did not fully understand, that did not preclude the possibility of the space being able to bring forth affective or discursive material with religious significance. If I did not find ways to deal with felt differences between myself and the bulk of people at the conference, the theo-political affinities I had imagined would just happen never would, because building community (especially one in which different religious and political praxis are present) requires the work of intervening in your own felt antipathies.

This realisation helped me to begin to question the legitimacy of my unexpected knee-jerk distaste for some of the ecumenical mixing that the conference convened. These reflexive conclusions helped me to apply a more responsive reflexivity when bumping up against ecumenical differences during the rest of the conference (e.g. reciting liturgy that contained what I perceived to be 'slack' theology) in an attempt to cut-in on my own reactionary responses to difference with a hopeful posture. When I say 'hopeful' I mean that without fully understanding the way people were practicing in the space, I tried to participate wholeheartedly, recognising that just because I did not 'get' what was going on, something significant could still come out of it. The reflexive work that the worship space prompted generated a significant change in my religious praxis during the conference (and beyond) and helped me to perceive practicing ritual alongside difference as a way of deconstructing sectarian divisions. By stumbling upon an excess of religious significance through sharing the peace, I had been forced into reflection. This reflection had led me to recognise some of my own internal contradictions and respond differently to the experiments in ecumenism that the conference convened. I deployed a micropolitics of hopefulness in foreign-feeling religious environments by opening up to deeply felt difference in order to see whether there might be opportunities to form a new and more nuanced theological praxis.

What we learn from Sutherland's experience of the conference is that religious landscapes are replete with different sources of religious significance and techniques of self. The religious person undergoes a complex process of drawing connections between different discourses, affects, and practices in order to generate their religious subjectivity. As Holloway (2003 and 2013) argues, religious people often repeat certain practices (like – in Sutherland's story – singing songs and sharing the peace) as techniques of self in order to alter their affective state or buttress their beliefs. However, these are also spaces of performance – and therefore uncertainty (Richardson 2013) – generating events of transcendent significance that might cut against that which is expected. Sutherland's account illustrates this when he discusses his unanticipated emotional response to sharing the peace, which in this instance could be seen as an effective technique of self, used to bond the group together. However, the story also highlights how the collective singing experience missed its affective mark, suggesting that collective techniques of self, organised in religious spaces, reveal themselves as an affective topology, comprising an uneven set of fluid responses to collectively engendered experience. Most importantly for this section, we observe receptive generosity unfolding in Sutherland's individual techniques of self, when he reflects upon the legitimacy of what is done in the space. Although prompted by the affective experience of the worship space and being swept-up into a postsecular emergence of feeling in-common, Sutherland's own time of reflection represents a kind of spiritual activism (McIntosh and Carmichael 2015) that reconstitutes his self-to-self and self-to-other relationships. He argues that this helped him to engage with the rest of the conference in a way that was open to the new possibilities for religious praxis that it might generate. As Sutherland (2016) makes clear, this transformed engagement included more generous encounters with a range of religio-political organisations involved in protesting the arms trade, environmentalism, and conflict resolution. The collective and individual techniques of self that the conference convened and prompted changed his perspective, enabling him to renegotiate his relationship to the religious and political organisations that he was already a part of. In particular, the conference gave him renewed conviction in holding his own church community to account regarding social justice issues. Reflection on ongoing, placed praxis was one of the stated aims of the conference, and Sutherland's story and our analysis of it illustrates how the conference fostered a receptive generosity that enables subjects to launch into more targeted political action based on a postsecular foundation of ethical negotiation and emergent senses of affect in-common.

Although this example shows how techniques of self can extend subjects' postsecular praxis into direct political action, it must be acknowledged that although there are examples of collectivity in postsecular political activism (see Williams *et al.* (2016) on the politicisation of foodbanks; Jamoul and Wills (2008) on community organising; Mavelli (2012) on the Egyptian Revolution), receptive generosity in postsecular moments does not always propagate a collectivised movement towards the political sphere. The ethical proximity and affective emergence that characterise postsecularity are often based on a recognition of commonality; of mutual vulnerability and shared desire. Receptive generosity is about building on the trust that these phenomena engender and enacting a further stage of relational transformation via interrogation of difference. Engaging in spaces of relational experimentation (such as participatory art, performance, or ritual) can create an atmosphere in which the tenets of postsecularity can emerge serendipitously. However, receptive generosity necessitates work. Reflection upon the legitimacy of friction in self-to-other relationships is at the core of receptive generosity, but in order to release some of these tensions, consideration of self-to-self connections is crucial. This kind of reflection enables changes to be made in the way that the subject relates to the Other but may also alter which Other the subject wants to engage. Reflection can precipitate realisations of where the subject's activist energy is best spent. This means that a group in which the participants are exercising receptive generosity – listening patiently to different perspectives and reflecting on their responses to them – may give rise to a profusion of differing political praxes rather than a collective one. In Sutherland's (2016) work, we can see an example of how this functions in his writing on his own community, Exeter Church (pseudonym), and how this gives rise to a further, more practical, manifestation of receptive generosity.

Exeter Church had been set up partially as a reaction to models of church that had left people feeling dislocated from their church community, unable to question theological hegemonies, and with little sense for how church activities had anything to do with the rest of their life. Against this, Exeter Church was to be a community based on mutual affirmation and dialogue concerning each others' faith praxis, democratic consultation on organisation, and a developing sense of how the church was changing the city. However, as Sutherland chatted to church members over the course of his time doing research with them, a theme that recurred was that there was little clarity over how the church was having an outward impact. This was despite the fact that the church had been involved in various activist projects. These projects included a subvertising campaign (fly-posting pro-hospitality slogans around the city at the height of a national right-wing media onslaught against refugees), piloting a lunch club at a primary school to combat holiday hunger, attending anti-austerity protests, collecting food and clothes to send to the Calais refugee camp, running fundraisers for refugee charities, and helping to run ecumenical services on 'Homelessness Sunday'.[6] Even though the church had been involved in a lot of activism, there was little sense amongst its members that it was having a noticeable impact on the city, or that relationships were emerging between the church and other groups that could build the capacities necessary to have a sustained activist impact.

Part of the reason for this diminished sense of impact was that the activist projects that the church had been involved in had largely been spearheaded by individuals within the community doing things that they were passionate about. These passion projects emerged as a result of connecting the reflection upon self-to-self and self-to-other relations that the community encouraged with praxis. The postsecular ethical proximity that the community fostered helped people to think through their individual praxes together, enabling the emergence of a hopeful affective topology. However, this sent people off on divergent paths in terms of how to convert this hopeful affect into action. To say there was no collectivity within this at all, however, would be a misrepresentation.

When people wanted to get behind a particular project, the rest of the church often rallied round, offering time and resources in order to help the person pull-off whatever action they were involved in. This way of organising, although lacking the continuity of an affinity group focussed on a singular issue (see Epstein 2002; Mellucci 1996), represented a practical manifestation of receptive generosity, where people not only listened to (particularly political) difference, but took a ride-along with it. Although the setting of reflection – or as we have suggested it might be called, spiritual activism (McIntosh and Carmichael 2015) – gave rise to a diffuse range of political praxes, practicing receptive generosity enabled members not only to listen to positionalities that they did not fully understand, but enabled them to provide the resources (at the expense of being distracted from their own passions) in order to bring other members' projects to fruition. By being willing to engage in political experiments that perhaps lay outside of many members' wheel-houses – strange territory for them to venture into, considering their own ethico-political predilections – receptive generosity extended the church's political activity and capacity into a plethora of different spheres. Receptive generosity has so far been conceptualised as a listening exercise (Coles 1997; 2001; Connolly 1999); experiencing discomfort by sitting with difficult or challenging perspectives and reflecting upon this discomfort. In this reflection, new rhetorical or discursive configurations can emerge. However, if groups practice alongside difference, this not only opens people up to reflection upon the action after it has happened, but immersion during the action into unknown affective topologies which transcend rationalist evaluation. Receptive generosity has the ability to plunge subjects into the uncertainty of performativity, assembling subjectivites, practices, discourses, and materiality in such novel ways as to allow new subjectivities and affects to emerge. This happened in Exeter Church when a more timid member supported some of the others by going to an anti-austerity protest with them. Although apprehensive that they would be intimidated by the protest environment and unsure that that kind of activism was for them, they found it to be a place of joy and of self-expression, helping them to feel more empowered and emboldened, extending their burgeoning political awareness and radicalising their praxis.

As the example of Exeter Church shows, although receptive generosity does not necessarily lead a group towards a collective affect that points in a unified political direction, a group can continue to practice receptive generosity in a way that extends its political capacity. Sutherland's account of the conferences also highlights how the listening practices of receptive generosity can build on the ethical and affective base of postsecularity, utilising reflexive techniques of self to deconstruct relational tensions. Receptive generosity can therefore be recognised as a significant way of extending postsecularity into more direct political action. The other way of extending postsecularity into the political realm is through rapprochement, which we turn our attention to now.

5.4 Postsecular rapprochement – the example of Occupy

In Chapter 1, we highlighted the significance of rapprochement as a way of building upon the ethical proximity and affect-in-common of postsecularity by creating 'crossover narratives' (Cloke and Beaumont 2013, 37): rhetorical constructions which enable a diverse community of actors to work towards common ends. Rapprochement generates crossover narratives because it is a process by which subjects with differing identities and political preferences co-produce movements and organisations through dialogue. The full slate of analyses and purposes that would characterise a subject's own ideological primacy is deferred in order to form resonant nexuses of political ideas, ends, and tactics in order to be 'doing something about something'. Rapprochement – when deployed as a conscious tactic – differs from receptive generosity because it focusses less on subjective transformation through encounters with the Other. Instead, it builds on the ethical proximity of postsecularity – found in the common affect of wanting to do something about something – by pragmatically reterritorialising political praxis and subjectivation across a range of hybrid spaces. This facilitates the co-production of bricolage-style political narratives with the Other in order to serve specific political ends that emerge from a common desire or impetus. This involves shrewd dialogue infused with the recognition that all sides will have to leave some of their disagreements 'at the door' and create a new political narrative that is a pastiche of their concerns and ideals and yet generates a multi-party resonance that enables proactive political co-participation. Rapprochement is a tactic chosen in order to get something done and is minimally concerned in its initial stages with the emergence of novel affective topologies. However, by drawing difference together to achieve a common goal, it is filled with the potential – that other mundane spaces of common endeavour, such as the artistic spaces we discussed earlier in this chapter – for the emergence of new affective topologies and subjectivities that are of political salience being that it is in the realm of political action that they emerge. The harder boundaries of agreeing upon a common political goal allow ideological, tactical, and motivational differences to be re-evaluated in an environment of reduced anxiety that these differences are going to impinge on the task at hand. This can mean that, over time, less stuff has to be 'left at the door' for the same group of people to work together and new, more developed crossover narratives can emerge. From these more developed narratives, new trajectories for affect-in-common, action, and identity can be realised.

In order to illustrate how rapprochement works, the quality of its political brand of postsecularity, and its political utility, we will draw on discussion by Cloke, Sutherland and Williams (2016) concerning the role of religion in the Occupy protests. Occupy was a moment of insurrection against bourgeois elites – the '1 per cent' opposed to the self-proclaimed 99 per cent of the Occupiers – stemming from common outrage across a vast panoply of different communities brought on by a neoliberal constellation of austerity and flimsy consequences for a profligate and socially irresponsible financial sector. However, within this commonality we are interested in the rapprochement – dialogue about differences and how they can be integrated or not to achieve common goals – between Christian occupiers and the rest of the movement. This may seem like an arbitrary division based on the heterogeneity of the movement, but we argue that distinctively Christian elements were particularly influential in the crossover narratives which defined Occupy. Christianity not only helped to characterise the content of Occupy's political action, adding to its heterogenous bricolage of political influences, but Occupy's engagement with Christian discourses, spaces, and subjects also created a more developed set of crossover narratives with even more wide-ranging political implications. As a result, Occupy became a moment as concerned with political theology, ecclesiology, and spirituality as it was with economic injustice. This led to a set of challenges from Christianity to Occupy and vice-versa, generating new kinds of questions and demands of both Christian and secular institutions that Occupy found itself in opposition to.

At a basic level, many Christian actors became involved in Occupy because of a shared felt intensity of the affective co-ordinates in which many people found themselves after the banking crisis of 2008. As a statement on the Judson Memorial Church website (2011) – an institution involved in Occupy Wall Street – said, Occupiers and a significant cohort of Christians within and beyond Occupy shared 'frustration with an unjust society … desire to speak truth to power and … hope that a better world is possible'. This represents the common affective baseline that is key for the emergence of postsecular generosity. It facilitated an intuitive openness or desire to search for justifications and resonances that would enable political collaboration across difference. This commonality enabled prominent spectacles, actions, and discourses that Occupy convened to be imbued with Christian resonances, partly as a display of tolerance towards these actors within the movement but also as a recognition that these discourses could increase the power, reach, and accessibility of Occupy's message. Additionally, this influence of Christian actors within Occupy enabled the movement to speak back in an apposite way to institutional Christian spaces that formed part of the Occupy assemblage – for example, St. Paul's in London and Trinity Church Wall Street in New York – in both hospitable and antagonistic roles. By connecting with Christianity in these ways, the momentum of Occupy increased, gathering resources, attaining more media coverage, and extending its scale. For example, in the USA, 'Occupy the Dream' was a movement – particularly within black churches – that amplified resonances between Christian theology and the message about injustice that Occupy foregrounded. This swelled numbers of-and-at protests and provided relationships between Occupy and churches, that lent the movement resources, providing places to meet, eat, and shelter, as well as access to a wider audience and contacts (Galindez 2012). In the UK, Christians that were part of the Occupy movement staged theologically inflected protests that highlighted the corporatist synergy between St. Paul's Cathedral and the financial district of London. Women dressed in white chained themselves to the pulpit during a service at the cathedral, demanding a greater indication that the church had a preference for the poor (following

Gutierrez 1988), and as part of the eviction of Occupy from the area around St. Paul's, the media spectacle included the removal of praying Christians from the cathedral's steps (Winter 2017; see also Christianity Uncut website).

Alongside the basic affective resonance that existed between some Christians and other participants in Occupy, and the obvious material and mediated resources that churches added to the movement, there are also two significant crossover narratives that arose from the co-productive role that Christians played within Occupy. First, there was the role they played in improving the prefigurative community of Occupy by providing (non-proselytising) spiritual counselling. A significant part of Occupy was not just protesting the injustice of the current order but prefiguring a pluralistic and horizontally democratic society in the encampments (Rieger and Pui-lan 2012). This helped to create crossover narratives about what kind of society that Occupy was trying to call into being. The introduction of 'protest chaplains' from a variety of spiritual traditions, including Christianity, generated an environment where 'people with deep spiritual struggles found a safer space to talk about those struggles' (McKanan 2011). The non-proselytising and pluralistic 'sacred spaces' set up by the multi-faith protest chaplains, stressed the importance of spirituality for society (and for social justice activism in particular), but also the necessity of spaces where spiritual traditions could exist together and learn from one another, so as to foreclose sectarian divisions (see Dittmer and Sturm 2010; Luz 2013). Second, Christianity added to the crossover narratives evident in the symbolism and spectacle of Occupy. For example, as part of Occupy Wall Street, one of the noticeable symbols of the protest was a gold papier-mâché bull with the words 'False Idol' written under it. This not only highlighted synergy between anti-capitalist polemic directed at Arturo Di Modica's 'Charging Bull' and the Judeo-Christian renouncement of the 'golden calf' (Exodus 32:4, ESV), but also highlighted that Occupy had a problem with both the material economic injustice of neoliberalism and its spiritually deleterious thinking that frames planetary wellbeing in purely economic terms (Eisenstein 2011; White 2018). This presence of spiritual thinking within Occupy – bolstered by a Christian presence and supported by Judeo-Christian symbolism and staging (see Ekkesia 2011; McClish 2009)[7] – helped to transform Occupy's message from being straightforwardly anti-capitalist, into a visionary and hopeful crossover narrative that not only suggested the end of capitalism but promoted a way of living marked by an awareness of interdependence and the consideration of meaningful lives.

As these examples have illustrated, Christians played a significant role in co-producing Occupy, leading to crossover narratives flavoured by both Christian and agnostic anti-capitalist concern generating a broad-based resonance, or at the very least, the ability to tolerate one another in resisting neoliberalism together. However, these initial crossover narratives enabled Occupiers to more consciously explore one another's differences, leading to more complex narratives and subjectivities arising from the movement. This generated challenging discourses that reverberated within and beyond Occupy, challenging both

Christians to think about their faith and its public role, and Occupy regarding their inclusion of spiritually inclined activists and activism.

Regarding the challenge to Christianity; as Christians worked participation in Occupy into their praxis, this showed the fluid nature of Christian subjectivity and practice as Christian subjects forged meaningful, subaltern theographies (Sutherland 2017). This meant that not only did the crossover narratives of Occupy illustrate generous acceptance of the profundity and usefulness of certain Christian ideas and practices, but it also issued challenges to churches that set themselves against Occupy. First, Occupy challenged churches to think about how they needlessly delimit sacred space. In a few instances, churches offered hospitality to Occupiers before withdrawing this support, sometimes on the grounds of the 'sanctity' of church spaces. This implication that Occupy somehow 'sullied' church spaces illustrated that churches were often concerned with maintaining their economic practices and traditions rather than offering succour to those who have been marginalised. As evidenced in the 'sacred spaces' set up by protest chaplains – that catered to the spiritual practices of people of all faiths and none – and Occupy LSX's Sermon on the Steps, Occupiers were more than happy to recognise that secular and religious space is co-produced. Occupy illustrated a desire to work out these blurry secular-religious co-productions in the service of representing the 99 per cent. Many churches refused to follow this narrative, relying on the imaginary of religious space as a silo that can be extended via colonisation. This silo imaginary set churches that espoused it against religious Occupiers, who adopted the idea that there is an ephemeral and mystical relationship between space and religious experience, generated by a plethora of interdependent actors (Holloway 2003). Second, Occupy challenged churches by criticising their top-down theology. Sparke (2013) argues that Occupy spurned top-down interpretation of social structures in favour of multiple local meanings, diverse global relations, and complex layers of activist space. Christians within the movement took inspiration from this and challenged theologies that depicted God as 'the boss who calls the shots and rewards religious shareholders' (Rieger 2012, 1). Instead they argued for theologies that took into account their experience of the divine through the relationships, affects, and actions of Occupy. This approach wrestles theology away from a top-down structure and queries establishment religion when it cosies up to the hegemonic governing institutions. It points out that there must be room in theology to grieve with the marginalised, which is a difficult task when simultaneously apologising for the status quo. Third, Occupy challenged the propensity of many churches to treat activism regarding social justice as optional, or an added extra to the 'real' business of Christianity. Occupy challenged churches to conduct a 'subversive remembering' (Bloomquist 2012) of the early church which was populated with marginalised people, knit them into a community, and drew on their experience of affirmation as God's children to nonviolently resist the powers which oppressed them. Pointing out that these practices have fallen out of the mainstream of Christian practice is to flip the accusation of mean-spiritedness directed at Occupy on its head (see Carey 2011).

The crossover narratives forged by Occupy challenged Christianity, however, the involvement of Christians and other religious people enabled Occupy's crossover narratives to develop in a way that challenged its own way of thinking about reaching out to and including religious activists. First, as established religious actors generally withdrew support for the movement, non-established religious people supported Occupy by providing pastoral care and additional impetus to speak truth to power. This challenged glib characterisations of religion as inherently entwined with the status quo and generated new trajectories and understandings of the potential for future activism across imagined religious/secular boundaries. Second, religious people within Occupy enhanced understandings of religion within the movement more broadly, enabling the regularly mediated narratives of Occupy to speak back pointedly to religious institutions, using its own language. This made political theology a key theme of the movement's crossover narratives, speaking into public debates about the role society wants and needs religious institutions to play. Third, the pluralistic spiritual spaces of Occupy enabled new forms of politically engaged spirituality to emerge. Novel, hybrid politico-spiritual identities could emerge, benefitting from anti-dogmatic spaces to explore religious practices and discourses, generating greater enjoyment within the movement by facilitating discussion and exploration of integrated spiritual and political practice. This, in turn, enabled the creation of different themes that characterised Occupy, beyond straightforward anti-capitalism towards ideas about how to create meaningful lives that spurn usury and capitalistic acquisitiveness in favour of a recognition of interdependence, care, ethics, and justice.

Through the example of Occupy, we can see the political utility of rapprochement, generating a way of converting a collective affective experience into practical political action. This has the benefit of bringing people together across difference and creating the space in which subjective transformations can occur, developing new trajectories for political thought, action, and partnership. In the instance of Occupy this postsecular politics was effective, by creating temporary autonomous zones that wrong-footed capitalist governance in order to offer people an experience of direct democracy and unity of purpose (Bretherton 2011). However, by carving out significant space for spiritual concerns within the movement, Occupy was able to turn its focus, not only upon the economic activism of protesting against neoliberalism, but the spiritual activism of prefiguratively wresting people from the draining psycho-spiritual affect of late capitalism (see Fisher 2009; McIntosh and Carmichael 2015). Against the disempowering material and affective constructs of neoliberal governance, Occupy posited the embrace of a rebel spirituality that taps into deep-seated desire, dreams, and hopes rather than selecting from a dog-eared neoliberal menu. This spirituality celebrates the interdependence which logics of individualism and consumption undermine and refuses to compromise with the oppression of neoliberalism anymore (Mentinis 2014).

5.5 Conclusions

In this chapter, we have given examples of three different modalities of how postsecularity affects politics and offered suggestions as to why it is a concept which is useful for political praxis. We have illustrated that the first and foundational postsecular modality builds on common affective experiences which are generated by shared participation in a political moment, or through mundane and creative spaces, generating ethical proximity and leading to new affective topologies. These new topologies can naturally be extended into political action such as in foodbank activism, hospitality, and participatory art in which ethical work directs people towards common political ends through a shift in affective topology. However, sometimes this needs to be facilitated, using provocative practices to get people to think through how their experience of ethics might have political implications, such as in the religious communities that practically implemented speculative ontologies.

Postsecularity can also be extended consciously into the political realm by the two other modalities we discussed; receptive generosity and rapprochement. Receptive generosity is a tactic of subjective transformation that uses techniques of self to progressively re-evaluate the legitimacy of antagonisms. This can enable subjects to engage with the Other based on an acknowledgement of the inconsistencies in their own ethico-political predilections. This is not so as to undermine their own political reasoning, but to create further opportunities to experiment in engaging with the Other, and to re-evaluate their praxis in light of these deliberate engagements with difference. This was illustrated through accounts of faith communities in which structures and discourses were promulgated so as to encourage members to conduct experiments in praxis alongside the Other.

Rapprochement was demonstrated to be a tactic of pragmatic dialogue less concerned with subjective transformation and more interested in facilitating crossover narratives that enabled resonance between different ethico-political subjectivities and therefore moving swiftly on to practically addressing select political issues. This often means that certain disagreements that the different parties involved could let divide them are 'left at the door' in order to do something about something. However, in this coming together to address a political issue, the boundaries that this focussed action creates enables participants to feel safe enough to evaluate the legitimacy of their disagreements because this does not have to interfere with the action that has already been agreed upon. This can lead to the evolution of the initial crossover narratives developed for the group to initially function, built on renewed appreciation of the Other. As these crossover narratives do their transformative work, spaces of rapprochement enable new political subjectivities, partnerships, and ideologies to emerge, generating greater political hybridity in progressive alliances and multiple trajectories for future political action. This was demonstrated through the example of the Occupy movement which initially brought together spiritual and anti-capitalist activists together in their common frustration with the injustices of neoliberal governance. Over time, the crossover narratives of Occupy developed as a result of dialogue between the spiritual and political elements within it, resulting in a holistic critique of neoliberalism and the positing of a democratic, pluralistic, and just future society, as well as new politico-spiritual subjectivities.

Given the arguments developed in this chapter we recognise three significant implications for research and praxis regarding the imbrication of postsecularity and politics. First, postsecularity can help us to understand and extend the practices of what Naomi Klein calls 'progressive coalitions' (2017). Taking the broad coalitions – formed of indigenous, environmentalist, religious, union, NGO, and veteran's groups – that protested against the DAPL pipeline at Standing Rock as a key example, Klein argues that powerful coalitions can be generated when people follow the postsecular pattern of coming together to forge ethical resonance based on mutual political affect. As White (2016) argues, this coming together necessitates finding ways to overcome preconceived ideological and tactical divisions, which requires interrogating the psycho-spiritual roots of our political predilections and airing them in dialogue with the Other. Klein argues that new understandings of this emotionally charged territory and its connection to politics is the only way to generate hope in the face of right-wing populism. What is more, bridging across a range of groups with differing concerns is key for generating a bricolage of crossover narratives that can provide an interdependent and holistic vision of society and programme of action that can defeat fascism through a groundswell of faith in the Other, disturbing societal affective topologies dominated by fear and ressentiment.

Postsecularity can help us to analyse what is effective in particular settings as people try to generate progressive middles. What tactic is most appropriate? Is it better to be non-interventionist and let new affective topologies 'organically' emerge, or is it better to try and encourage collaboration through rapprochement or persuading people of the benefit of an alternative political ontology? In different settings, different expressions of postsecularity may be appropriate, and in certain settings attempts to forge postsecularity may even be inappropriate. However, by recognising the benefit that a postsecular ethos can bring to politics, analysts can offer a new way of appraising a political moment and suggesting new potential actions that search for progressive middles, when appropriate or useful. We argue that this kind of postsecularity evidences an implementation something akin to White's 'unified theory of revolution' (2016, 69) which recognises the strengths of different ideological approaches to activism and yet advocates that no one theory can provide the template for action, suggesting instead a pragmatic mix-and-match approach that listens to difference and discerns the signs of the times.

White's approach leads us to our second implication for research and praxis. When ideological stringency is called into question, new ontological approaches to politics can be entertained. As Mentinis (2014) argues in the case of the Zapatista movement, a new political ontology was embraced by the Zapatistas that moved past victimisation by the Right and identification as reliable voters by the Left. Similarly, Sullivan (2005) argues that for another world to become possible, political subjects need to move past the political status quo of secular parliamentarianism and embrace a new way that speaks to liberated desire. We argue that postsecularity is a way for subjects to explore new ontological approaches to politics which – following Connolly's (1999) reading of Nietzsche (1974) – recognises the fragile texture of ethical predilections and moves towards a subjectivity that is willing to tarry a little longer with difference (see Anzaldúa 2012). Although this is taking postsecularism to its extremity, we argue that it can lead to a new type of political consciousness like the Zapatistas' (Mentinis 2014) or Anzaldúa's Mestiza.[8] It is a consciousness that is willing to dislocate from the current symbolic order in order to channel something more enchanting and intangible, allowing the subject to imagine alternative futures that include ideological outliers and grey areas. Although the Zapatistas and Anzaldúa are firmly rooted in the left-wing political lexicon, how might recognition of the mystical elements of their ontologies be identified in other groups and ideologies and interrogated for their political implications and seen as political partners? Using a postsecular lens, we can more effectively blend geographies of politics and religion as our illustrations of Moody (2012) and Ganiel's (2006) work on speculative ontologies demonstrate. It is at these interstices of politics, religion, and spirituality that some of the dimensions of enchanted secular ontology that generate surpluses of meaning and affect in assemblages of the public (as reflected upon in Chapter 1) comes into play.

Third, we recognise that the arguments presented in this chapter generate new possibilities for participatory action research in human geography. Postsecularity creates a new way of assessing the political terrain that is built on finding crossovers and generating both openings to novel subjectivities and broad bricolaged resonances across difference. What new types of political tactics and actions can be dreamt up? And what are the new ontologies and end goals we can base these on? In this chapter we have examined a grounding of speculative ontologies through arts practice, faith communities that support their inherent political diversity, and new forms of hospitality built on increasingly broad networks of mutual aid. What new kinds of activism can be generated by conversations that blend politics and spirituality and that focus on bridging between deeply felt desire or pain and ethics? Postsecularity inspires us to dream up new ways of being together, and to address the question: how can geographers extend their participant activist research to begin to blend spirituality, politics, and ethics more consciously?

Notes

1 This identification of the parallels between Taylor (1987), and Deleuze and Guattari (1994) is drawn from Holloway's (2011) chapter on spiritual geographies.
2 Eagleton (2011) suggests this opening to 'raw possibility' is a kind of ressentiment that problematically suggests that progress can be made without ethical negotiation. This is an ego-move that tries to cover over the subject's weakness and fear; their inability to find the energy to engage in the difficult work of ethics or their worry that however much ethical work is put in, nothing will change for the better. Eagleton (2011) argues

that refusing to admit our limitations in the sometimes-terrifying futility of ethical work and reliance on the emergence of an ethically justifiable event is a kind of evil.

3 This is as opposed to conceptualising postchristendom as the phenomenon or social reality which Murray (2011) defines as 'the culture that emerges as the Christian faith loses coherence within a society that has been definitively shaped by the Christian story and as the institutions that have developed to express Christian convictions decline in influence' (p. 19).

4 For an in-depth discussion of these processes, see Sutherland 2017.

5 For discussion of how these types of knowledge affect the construction of theology see Beckford 1998; Gutiérrez 1988; Holland 2015; Rohr 2003; Talvacchia 2014; Williams 2013.

6 For explanation of 'Homelessness Sunday', see Housing Justice 2018.

7 Using Christian symbolism and staging as part of crossover narratives in pluralistic movements is not unique to Occupy. See Sutherland's (2016) work on working class protests in the UK's Midlands against the 'crucifixion' of local libraries.

8 See also the political role of Rastafarianism and Peace Concerts in Marlon James' (2014) novel *A Brief History of Seven Killings*.

6   Wider religious and spatial conditions

6.1 Introduction

This chapter critically interrogates the geo-spatial limits of postsecularity in terms of its empirical study, which, up to now, has tended to privilege Christian-secular nexuses of engagement. We assess the ways in which distinct spatial contexts shape both the emergence and constitutive dynamics of postsecularity. This approach helps map how postsecularity is co-constructed as a series of affective, spatial, and policy practices across non-Western social-spatial and historical contingencies. This provides a more nuanced analysis of the heterodox ways receptive generosity, rapprochement/partnership, and re-enchantment are being generated globally.

For definitional clarity, it is important to note that we make a distinction between the postsecular and emergent postsecularity. While the former concerns the co-constitution of the religious and the secular in modern life; our formulation of postsecularity refers to a way of being that arises from faith and secular motivated actors relating differently to alterity, a process that is marked by the 'crossing over' of religious and secular narratives, practices, and performance, which results in the re-territorialisation of belief, identity, and ethical praxis.

Formations of postsecularity are intimately entangled with place, identity, and culture, underlining the importance for scholarship to assess the different types of secularism and the variegated geographies of secularity, religion, and (un) belief that shape and circumscribe the proclivities for postsecularity in different places. In what follows, we critically modulate elements of our working definition of postsecularity across newly emerging spaces, subjectivities, and movements as they are appearing in key literatures and research agendas. These include debates surrounding gender and urban/public space; the changing perception in international development and humanitarian studies of the traditional relationship between the religious and the secular with regard to notions of development and progressive change; new understandings of the State/religion nexus generated by the intervention of postsecular thinking, chiefly in relation to China, Russia, and India; and the perspectives of assemblage and actor network theory which sees numerous non-Western and non-Christian sites being ethnographically explored in order to expand our understanding of the ways that 'the religious' reproduces urban modernities, and vice versa. In mapping out this terrain, we seek to emphasise the highly contingent and localised formation of, and barriers to, postsecularity

6.2 Gender and postsecularity

The exclusion of women's bodies and experience from both public and urban space has been a growing subject within postsecular debates (Greed 2011; Graham 2015). Within the West, Greed for example, has clearly delineated the patriarchal power that has been historically wielded within both planning and cultural assumptions that link the public space with male agency and activity whilst female experience is 'contained' either within special planning zones, or within the domestic arena (Greed 1994 and 2011). The secular planning assumptions around a gendered public/private split within urban life is often reinforced by religious and cultural norms whereby women and young people are doubly excluded because of their gender – invisible and inconsequential to the secular eye and often vigorously policed by the religious one. This invisibility generally frustrates women who take on board public positions and roles of leadership within both the worshipping and wider community (Greed 2011, 113). At the heart of this dynamic, Greed says, is the need to regulate female sexuality and reproductive power for the sake of a male self-image predicated on control and rationality. Some feminists are therefore expecting the shift from a secular to a postsecular discourse to have a liberating impact on the role and expectations of religious women to participate more fully in public urbanity. Elaine Graham, for example, cites the work of Luce Irigaray, Julia Kristeva, Donna Haraway, Judith Butler, Rosi Braidotti, and in particular, Grace Jantzen, as examples of feminist philosophy that engage creatively with religious ideas and spirituality as conditions for a new public modernity (Graham 2013, 58–59).

However, the category of the postsecular is, in Graham's view, still an ambivalent category that heightens a tendency for secularism to acquire harder confrontational edges such as the positionalities expressed in so-called New Atheism. A sometimes intolerant liberal and feminist left continues to fulminate on issues such as the public wearing of the veil in Islam, whilst from the religious right, women's bodies continue to be contested sites around issues such as abortion and reproduction. Graham concludes:

> For many women around the world … the post-secular does seem to leave them between a 'rock and a hard place': between the global resurgence of religion and multi-cultural appeals to difference and tolerance, and the imperative to protect the well-being and self-determination of women and girls in the face of authoritarian theologies.
>
> (2013, 60)

On the more optimistic side, some feminists argue that the concept of the post-secular does open up new spaces for valourising spiritual and religious sources of agency. This creates a challenging caesura to the widely held assumption in Western liberal strands of feminism, that religious identity and practice represents only a source of repression for women's liberation and expression (Moghadam 1995). Braidotti (2008), for example, argues that the postsecular category broadens existing understandings within secular critical theory by insisting that spirituality and religious piety are legitimate forms of agency and subjectivity (p. 15). In research terms, it prioritises the 'multiple micro-political practices of daily activism of interventions in and on the world, we inhabit' as 'strategies of affirmation' (p. 16). She concludes with a definition of postsecular subjectivity 'as the ethics of becoming: the quest for new creative alternatives and sustainable futures' (p. 16).

From an Islamic perspective, the work of Saba Mahmood and her book *The Politics of Piety: the Islamic Revival and the Feminist Subject* (2005) unambiguously challenges the assumptions of liberal and postmodern feminism. Her theoretical standpoint emerges from the Islamic revivalism associated with the Egyptian mosque movement, which emerged as a reaction to the secularisation and westernisation of post-colonial Egypt. Her stance seeks to directly disrupt the hegemonic nexus within Western feminism and critical theory that true womanhood can only reside in notions of autonomy and agency that axiomatically act in opposition to transcend what she identifies as the 'constraints' of tradition, custom, and of transcendental will (p. 8). Mahmood deconstructs this hegemony by redefining the terms of engagement for those who choose to operate outside this paradigm, by staying in those cultural and political spaces that 'are not necessarily encapsulated by the narrative of subversion and the inscription of norms' (p. 9). For example, her sample of Muslim women engaged in the Egyptian mosque movement cultivate Islamic notions of docility, piety, and the submission to an external authority as an expression of 'positive ethics' that contributes towards the reaching of one's true potential. In Islamic ethics, the Western disjunction between desire and social norms is often collapsed. They are not seen as mutually antagonistic, but mutually affirmative. As Mahmood writes,

> To conceive of individual freedom in a context where the distinction between the subject's own desires and socially prescribed performances cannot be easily presumed, and where submission to certain forms of external authority is a condition for achieving the subject's potentiality.
>
> (p. 31)

This ambivalent stance to theoretical postsecularism also gets played out in the way feminist studies of spatial practices associated with women and religion in non-Western contexts are interpreted; for example, in the renewed visibility of the headscarf in Turkey (Gökarıksel and Secor 2015). This visibility is linked to wider geo-political and economic issues, not least the growing wealth of an Islamic middle class and their desire to create gated communities that allow for the expression of a more overt Islamic identity. It is also linked to a wider resurgence in Turkish nationalism, in which appeals to the Neo-Ottoman heyday of Turkish colonialism and global prowess under its Islamic guise, jostle and threaten to overturn the more recent but perhaps less deeply-rooted secular vision of Turkish laicite (or *laiklik*), based on the French Republican model, and promulgated after the collapse of the Ottoman Empire by Kemal Ataturk in the 1920s.

Nilufewr Göle (2010) in her analysis of Turkish *laiklik* places young womens' search for both a modern and Islamic identity within a symbiotic reading of modernity, secularism, and religion. At the heart of her analysis is the idea that the power of the secular lives on in the way that religious citizens subscribe to a 'voluntary secularism' (p. 248) that involves the presentation of a 'civilised self' within the phenomenology of everyday life.

Central to this thesis is the way that macro processes of secularism 'discipline' micro-practices at the individual level. According to Göle, Turkish secularism invokes a set of interdisciplinary practices that operate at the level of state, law-making, and constitutional principles that establish a 'rhythm' to the phenomenology of everyday life practices, and thus becomes embodied in people's 'agencies and imaginaries' (p. 254). Göle then deploys the framework of Bourdieu's habitus to describe how these embodied agencies and imaginaries form a secular habitus, which is considered to be a higher form of civic belonging. This secularism becomes a learned practice to be performed by new elites who were educated at influential republican (i.e. secular) schools and where, for example, Arabic and Persian elements are eliminated from the modern Turkish language curriculum on account of their theocratic pedigree. This secular habitus is further reinforced by the construction of privileged secular urban spaces, to rival the influence of religious ones, including opera and ballet houses, concert halls, and ballrooms in which new social and culturally acceptable ways of interacting (such as men and women holding hands in public) are promulgated (p. 255).

The emergence of Islam in the post-1980 period, challenges this hegemonic control of the secular, and is especially felt by female Muslim students who now decide to wear headscarves on university campuses. According to Göle, it is both religious visibility and femininity that defies the hegemony of the secular over definitions of self, sexuality, and space. But of course, this new and public symbol of religious identity and piety also challenges *religious* hegemonic assumptions and practices, by breaking the normative taboos and assumptions of women wishing to claim intellectual and public agency in 'secular space'. This desire to have 'access to secular education follows new life trajectories that are not in conformity with traditional gender roles and yet fashion and assert a new pious self' (p. 255). With regard to the headscarf, the stigma, and inferiority traditionally associated with it within some secular discourses, instead becomes imbued with values of distinction and prestige, and the confident acquiring of cultural and social capital. Thanks to the booming fashion industry now associated with headscarves and other traditional Islamic clothing, the development of a new aesthetics for Muslim women is beginning to emerge (Tarlo 2007).

It is important to understand how these studies illuminate the conditions of postsecularity, and in particular around our ideas of the emergence of new ethical and political subjectivities. For example, the shift towards a consumerist, individualised young Muslim woman's identity in which trajectories of self-fulfilment appear to represent a blow to patriarchal and other 'outdated' religious norms would seem to be a decisive shift in favour of a secularised hegemony. Within this narrative, the headscarf issue in Turkey is interpreted as further evidence of the power of a differentiated public space (Casanova 1994; Riesebrodt 2014), whereby religious identity becomes increasingly hollowed out and deracinated from traditional Islamic cultural and scriptural norms. Younger Muslim women's aspiration to a secular university education and other 'secular benefits' suggests they are creating their own versions of Republican elitism (Göle 2010, 255). However, this apparent narrative of the secularised isomorphism of religion is counterbalanced by examples of how secular Turkish and pro-feminist groups seek to mirror the practices of religious communities in the form of public marches and protest movements on the streets (Özyürek 2006). Göle describes them thus:

> The secular formed a mise-en-scene by masses getting together, by symbols (oversised flags acquiring a new popularity), by photographs and sayings of Ataturk, but also by new modes of secular clothing for women (in tune with the colours of the Turkish flag): red and white miniskirts, ties and caps ... Secularism was performed in the public collectively and visually ... a new market in icons and clothing attempted to create a secular fashion and the use of Ataturk's pictures, deeds and words provided a frame for commemorating Turkish secularism.
>
> (p. 259)

Rather than mono-directional isomorphism towards either the religious or the secular, a 'geographies of postsecularity' reading of the public space could be said to affirm a 'mirroring', whereby religious and secular practices and imaginaries both influence each other so that the public space is now changed into a fluid arena characterised by multiple expressions of agency and redistributions of power that mitigate against any definitive reading. Göle expands this thesis. 'Muslim women cover their bodies yet become more visible to the public eye and hence unsettle the religious norms of modesty and the secular definition of the feminist self' (p. 257).

However, this new hybridised identity that combines both 'piousness and publicness' is not one that emerges through chance or through an eclectic agglomeration of social actants (as in some readings of assemblage theory – for example, Storper and Scott 2016). Rather, it is a conscious and dynamic shaping of a new ethical subjectivity and identity that conspicuously seeks to refashion both the religious and secular habitus, even if the precise destination of this refashioning is yet to be defined. Göle ascribes it (after Victor Turner) as a 'performative reflexivity' in which 'perceptive members of a socio-cultural group turn back upon themselves, upon relations, actions, symbols, meanings, and codes ... to make up their public selves' (p. 263). According to Göle's reading (after Taylor, Turner, Foucault, and Bourdieu), Islam is shaped by the secular age, but in doing so it brings forth alternative notions of self, morality, and piousness that in turn publicly shape and internally re-arrange the habitus of the secular. This more contested and edgy reading of gendered postsecularity, warns against an over-smooth or unproblematised overlaying of the postsecular over religious identities. If there is a new sense of agency emerging for Muslim women in this space, it is unclear if this is at the expense of their traditional religious heritage and identity, or whether it will generate new forms of solidarity, resilience, and empowered identity.

Rosa Vasiliki is sceptical on this latter point (2016). She agrees that the 'postsecular turn' (McLennan 2010) in feminism has exposed many hidden hegemonic and outdated secular assumptions about women's agency and autonomy. However, the postsecular move towards affective and ethical engagements nevertheless creates a dangerous shift away from using and developing critical theory itself. Postsecular feminism, far from being an emancipatory dynamic for non-Western women, is in danger of simply becoming apolitical. An account of agency, devoid of the importance of separating from cultural norms (understood within feminist studies as autonomy) simply reinforces the status quo because it fails to adequately engage with 'questions of inequalities and the ideologies that sustain their natural status' (2015, 119).

There is probably now sufficient empirical research on Muslim women's engagement in political struggles in both the UK and beyond to challenge Vasilaki's 'apolitical' critique. Politically motivated events such as the Arab Spring of 2011, and the Prevent agenda in the UK have been catalysts for a more explicit and assertive politicisation of affective and ethical stances within Muslim women. For example, Lewicki and O'Toole (2016) analyse the role of grassroots Muslim women initiatives in raising policy agendas and practice regarding FGM in diasporic Islamic communities and lobbying in favour of mosques giving equal access to women's participation and leadership. More recently, Muslim women from diasporic communities in Wales have been active challenging the instrusive, disproportionate, and highly racialised safeguarding checks surrounding FGM (Hibbard 2018). Meanwhile, Ahmad and Rae (2015) discuss the still largely unreported role of both secular and religious women in the occupation of Tahrir Square in the early days of the revolution in January 2011, before the Army and other men began their heinous policy of search, rape, and intimidation (McBroome 2013). Ahmad and Rae identify other areas of Muslim women's engagement in the public square:

> Women in countries like Iran, Malaysia, Pakistan, and Turkey are negotiating new notions of an Islamic public square, perhaps wearing a veil and conforming to Islamic protocol and are thus allowed to be in public and participate in that capacity, sometimes making specific demands related to marriage laws associated with age of consent, inheritance, and property

rights that comply with Islamic norms. Now, many Muslim women proudly don the hijab as a sign of resistance to Western policies and Islam is not seen as a symbol of servitude, but of common ground and a rallying cry to unite citizens. For women in particular, they are finding a space within the everchanging Middle Eastern landscape; and for many within a reformist Islam where women had the right, the *khilafah*, to demand change, it was embraced when the Arab Spring began to flower.

(p. 316)

Prominent young Muslim women like 2014 Nobel Peace Prize winner, Malala Yousafzai, who, as a teenage school girl survived a gun attack on her by the Pakistani Taliban in 2012, are strong advocates for an equal education system for girls and young women. However, Ahmad and Rae are also quick to point out, that the gains in women's rights made in the early days of the Arab Spring remain fragile and vulnerable in the face of often violent retrenchment by both patriarchal and nationalist factions (p. 318).

This section, exploring how the notion of the postsecular critically intersects with non-Western readings of feminism, elaborates our theoretical work on post-secularity in the following ways. First is the notion of new subjectivities created by the way the postsecular interrupts normative Western readings of women and religion. These subjectivities present alternative narratives of becoming, which redefine our understanding of female autonomy and agency in public and urban space. These subjectivities, and the ethics that accompany them, are predicated on notions of reaching one's potential through the connected webs of culture, religion, and family rather than an uncritical and solipsistic transcendence of norms for its own sake. It is the ethics of connection and solidarity that recreate these alternative notions of agency and autonomy.

There is a suggestion that Western secularism still dominates individual sub-jectivities and continues to secularise traditional religious structures of authority and culture (such as in the modest fashion debate concerning the public wearing of head coverings). However, Islamic culture and fashion are also re-defining secular notions of 'being' in the public square. So rather than a one-way iso-morphism narrative traditionally favoured by normative theories of secularisa-tion, a more likely reading is that the iso-morphistic pressures generated by postsecularity work both ways, moulding the secular and religious into new hybrid forms of expression and practice. Finally, in line with the evidence from our case studies, ethical impulses to engage in geographies of postsecularity are increasingly leading to political conscientisation and new forms of rapproche-ment between secular and religious Muslim women engaged in health and education issues that directly affect women and young girls.

6.3 Postsecularity and international development

Another literature where there is a growing discourse around religion and belief, and notions of post-colonial agency is in the field of international development studies. Religion in international development, particularly the role of faith-based organisations, has received growing policy and academic attention in recent years (Deneulin 2009; Haynes 2007; Haar 2011; Clarke 2006; Oslon 2008; Atia 2012; Carbonnier 2013). The enduring presence of religious narratives and actors playing prominent roles in humanitarianism has led scholars to re-examine some of the received assumptions of secularisation and secularism (Deneulin and Rakodi 2011; Ager and Ager 2011). Previously, religion had been regarded as irrelevant in the quest of modern societies to deliver economic stability (Rakodi 2007) and was marginalised under the functional secularism of both postcolonial governments and academic cultures alike. However, religion's continued presence in social and political life in developing countries (Jenkins 2007) has led to a recalibration of development discourse that pays more attention to the significance of religious belief and practice to contemporary global challenges. This recalibration also highlights the multiplicity of religious and spiritual interpretations of social, economic, and political realities, including polytheistic and non-theist sources (Deneulin and Rakodi 2011). Scholars have debated whether this marks a sea-change in the functional secularism of development discourse, or whether it resembles a humanitarian pragmatism (e.g. the conference on Religion and Human Rights Pragmatism: Strategies for promoting rights through dialogue across religions and cultures held at Columbia University in 2011).[1] Either way, very little is known of how these dynamics are worked out empirically in different donor to recipient settings.

Perhaps one of the defining characteristics, and challenges, in contemporary international development is the growing trend towards development consortiums and partnership (Deneulin 2009). Partnerships across the faith-secular and interfaith boundaries raise important questions for Clarke (2006, 846) who writes:

> The challenge posed by the convergence of faith and development is to engage with faith discourses and associated organisations, which seem counter-developmental or culturally exotic to secular and technocratic worldviews, in building the complex multi-stakeholder partnership increasingly central to the fight against global poverty.

Postsecularity provides a potential frame to analyse both the formation of these emergent collaborations as much as the barriers towards partnership. The never-fully-past history of colonialism, imperial control, cultural assimilation, and forced conversion linger in collective memory, leading to suspicion towards another faith group or organisation. Establishing the theological value and best practice of interfaith collaboration is fundamentally important according to practitioners in the field: 'We can only end world poverty by working with people who have faiths other than our own – so let's do more of it, dispel the myths and share examples of best practice', said Christian Aid's Inter-Community Initiatives Manager, Nigel Varndell (cited in Department for International Development 2009, 7).

Investigating the postsecular, then, differs from previous research which has predominantly looked at the operation of leading FBOs in the field (CAFOD, World Vision, Christian Aid, Progressio). The focus here is on collaborative campaigns and event spaces in order to ascertain the different ways narratives and practices can be created that hold shared ethical values across faith-secular and interfaith boundaries. The importance of postsecular ethics and crossover narratives is heightened in areas of development, given historical and contemporary conflicts stratified on religious and ethnic lines (Clarke 2006).

In the *Keeping Faith in Development* report, Kessler and Arkush (2008) present three case studies to encapsulate this theological and pragmatic move towards postsecular partnerships: Christian Aid's peace building work in Mindanao in the Philippines where there is a Christian majority and a Muslim and non-Muslim minority; Islamic Relief's work during and after the civil war with Christians and Muslims in Southern Sudan; and World Jewish Relief's work with a local Muslim partner agency in Kashmir, Pakistan after the earthquake. Research to date has under-theorised the ethical concepts underpinning interfaith and faith-secular collaboration and understanding the philosophies and practical workings of postsecular partnership is essential for best practice.

To do this, empirical examination is needed to assess the impact of partnership on the organisational practices of development in secular organisations and FBOs active in international development. Theological and developmental values endorsed by disparate groups are likely to undergo reformulation through closer collaboration. Postsecularity will be dialectically produced through practice and networks and will potentially affect donor mobilisation and delivery level differently. Key research questions that have yet to be fully addressed would include; What practical forms does rapprochement take? How is it initiated and sustained? How are potential differences within postsecular partnerships negotiated and resolved? What crossover narratives are put in place, and what are the boundaries of the scope of agreement? What ethical and pedagogical processes are at work that allow different social actors to recognise (but not necessarily endorse) the worldviews and methodologies of others? What can postsecular rapprochement achieve?

6.4 Postsecularity and the state/religion interface

A further dimension shaping the conditions for postsecularity concerns the different ways the newly visible interface between modernity, religion and belief, and the secular is played out within the era of globalisation and the perceived weakness of the nation state (Brinkman and Brinkman 2008; Suter 2008; Holton 1998). Classic models of state sponsored secularism that emerged in the eighteenth to the twentieth centuries usually refer to abstracted typologies of secularism. There is general agreement that there are four main types or models (Stepan 2011).

The first is *separatist secularism*, with a spectrum of intervention that includes separatist autocracy, authoritarian secularism, and fundamentalist secularism and which would include countries such as France, Turkey, Egypt, and Syria (Stepan 2011, 122).

Next is the *established religion model*, characteristic of many twentieth century Protestant Western European democracies including the UK, Finland, Sweden, Norway, Iceland, and Denmark. These are generally inclusive democracies in which the rights of religious minorities and majorities are respected. This view of the equilibrium inherent in this model is largely predicated on the combination of 'religious homogeneity' (e.g. Lutheran, Anglican) and decreasing intensity of religious practices that 'make it easier for an established religion and an inclusive democracy to co-exist' (p. 122). When the religious homogeneity is eroded by successive and sharp waves of migration and globalised interaction, which brings with it a concomitant rise in religious intensity, then the state-society-religion formula is put under pressure and changes can occur. The established religion model, however, seems at present a fairly robust and resilient one by which to absorb increased heterogeneity and intensity of religious activity, despite considerable political and cultural provocations and backlashes against further migration, and integration of religious and ethnic minorities.

The third model, the *positive accommodation* model, characterises many Roman Catholic but also multi-confessional shaped western European countries such as Germany, Switzerland, Belgium, and the Netherlands. It is historically linked to previous models of managing conflict between Christian religions. Under these constitutional norms, the state is 'obligated' to actively support religion including such areas as planning law and state education. Under this arrangement, for example, private Calvinist and Catholic schools in the Netherlands would also be entitled to receive state support.

The final model is defined as the *Respect All, Positive Cooperation, Principled Distance* model which is characteristic of post-colonial governments across different religious-based polities including India, Indonesia, and Senegal, all countries that have large Muslim populations. Their post-independence leaders instinctively rejected the US 'freedom of religion from the State' model of secularism. Sometimes called 'the principled distance model' (Bhargava 2007), it is conceptualised as a non-neutral and equidistant variant which says that if Religion A is violating citizens' rights, but Religion B is not, then the state will invoke its democratic powers against Religion A and not Religion B. It is predicated on a model of Indian secularism that accepts a nation characterised by religious pluralism and governed by a non-discriminatory state. However, as TN Madan points out, when it comes to the Indian example, the government has on many occasions 'ended up tying itself in knots, trapped in uncomfortable proximity instead of maintaining a reasonable distance (2011, 12).

However, as is now clear from evidence presented elsewhere in this volume, the intense cultural, political, and economic shifts of the past 20 years, generated by the constantly refreshing of globalised webs of impact and interaction, has radically destabilised these existing typologies of religion/state interface. We need to move beyond static and abstract definitions, and instead explore more fluid and contingent relationships. There is a growing literature exploring different concepts of the postsecular beyond the West, which attests to its growing global valency. This valency goes some way towards answering the critique that the postsecular is a redundant and colonial theoretical framework which underplays the ways the sacred and secular have always interacted in interesting, complex, and dynamic ways across different global contexts (Kong 2010).

We focus on three representative geo-political regions beyond the West where postsecular analyses are beginning to emerge; namely Russia, India, and China. These large countries are struggling to adapt to the rapid social shifts associated with globalised modernity. Once-normative narratives of culture and identity, initiated from the centre, and designed to hold fractious and heterodox cultures and communities together, are now fragmenting under the pressures of both internal and external change. Religion and belief are integral to these pressures, acting as both catalysts for disruption, but also as the social and cultural valve that absorbs these body blows and relieves the pressure in a way that allows change and transition to occur in a relatively smooth and enabling way.

6.5 Russia: desecularisation and oppositional postsecularity

The collapse of the Soviet Union in 1989 was for many the harbinger of the new post-secular age that began to emerge in the last two decades of the twentieth century (Caryl 2013; Jurgensmeyer *et al.* 2015). Since that time, reassembling a core Russian identity out of the ashes of the Soviet Federation has led to the inevitable rise of the power of the Russian Orthodox Church (ROC) as the state-sanctioned vehicle for a Russian nationalist and cultural identity that seeks to evoke the former glories of the Tsarist era (Karpov 2010). Legislation passed in 1997 on Freedom of Conscience and on Religious Associations has sought to curtail the influence of non-Orthodox religions with strict rules on legal constitutions, liturgy, and the banning of foreign leaders (Daniel and Marsh 2008). On the back of the cultural resurgence of Russian Orthodoxy, other religious groups and activities have been quick to occupy the new religious space in Russian public life. These groups include Pentecostal churches that are both locally but also internationally supported, as well as Muslim communities, particularly in those provinces of the vast Russian hinterlands where they are the majority (Kuznetsova and Round 2014).

Giudice usefully tracks the contours of what she calls desecularisation of the Russian Federation (see Karpov 2013). 'Desecularisation from above' refers to the legal and cultural homogeneity of the ROC. However, an expanding Protestant component, comprising of several Pentecostal churches and movements, is contributing to what she calls a 'desecularisation from below'. They achieve this via engagement in multiple sites of welfare and service provision which also include the freedom to unambiguously share the faith-based motivations for their work. The postsecular debate in Russia is therefore developing along two vectors. The first is the contested and ambiguous role of the Russian Orthodox Church as the upholder and gatekeeper of what is still constitutionally a secular Russian State. The second is a more granular postsecularity, based on the now ubiquitous role of religious groups and faith-based organisations in providing the local civic and welfare services that the Russian State is unable or unwilling to organise. This combined with low levels of trust towards the secular authorities is creating new and interesting examples of postsecular rapprochement.

Pussy Riot and oppositional postsecularism

One event which captured the world headlines in 2012, and still resonates across multiple settings to this day, was the guerrilla performance in Moscow's Cathedral of Christ the Saviour by the agit-prop music group Pussy Riot. The group entered the church dressed as regular visitors before removing their coats to reveal multi-coloured dresses and donning balaclavas to perform a so-called 'Punk Prayer' on the raised platform in front of the doors to the iconostasis. The prayer was called 'Mother of God, Banish Putin' and included the following lyrics:

> Virgin Mary, Mother of God, banish Putin
> Banish Putin, Banish Putin!
> Congregations genuflect
> Black robes brag, golden epaulettes
> Freedom's phantom's gone to heaven
> Gay Pride's chained and in detention
> The head of the KGB, their chief saint
> Leads protesters to prison under escort
> Don't upset His Saintship, ladies
> Stick to making love and babies
> Crap, crap, this godliness crap!
> Crap, crap, this holiness crap!
> [Chorus]
> Virgin Mary, Mother of God
> Become a feminist, we pray thee
> Become a feminist, we pray thee.

Three members of the Pussy Riot collective, Nadezha Tolokonnikova, Maria Alyokhina, and Ekaterina Samutsevich were arrested and sentenced to two years in a minimum-security prison on charges of hooliganism, although Samutsevish was later released on appeal. In a forensic and detailed exposition of the case, Dmitry Uzlaner (2014) highlights several key features of postsecularisation relevant to our thesis. First, the question of intent. Was the Punk Prayer a genuine, if highly untraditional prayer service, an intentional expression of blasphemy and sedition, an inappropriate artistic performance, or a strategic act of civil and political protest? (p. 27) Perhaps the easiest way to answer this conundrum was to suggest it was all four, and intentionally so. A Pussy Riot blog released at the time of the arrest stated:

In all our public statements, we constantly emphasise that the punk prayer 'Mother of God, Banish Putin' was truly a prayer – a radical prayer directed to the Mother of God to prevail upon the earthly authorities and ecclesiastical authorities who take their cue from them. Among the two dozen Pussy Riot members, many are Orthodox believers for whom a church is a deep place of prayer ... we did not desecrate the church nor did we blaspheme.... We passionately prayed to the Mother of God, asking her to give us all the strength against our incredibly merciless and wicked overlords. And we will continue to sing songs and will pray for those who want us killed and thrown into prison, because Christ teaches us not to wish death or prison on those whom we do not understand.

(quoted Uzlaner p. 28)

Other blog posts and utterances have also couched the event as a piece of political performance art, which chimes in more with the interpretation of the international campaign in support of the three plaintiffs.

Then there is the issue of *interpretation* of the event. The positioning of the members of Pussy Riot as believers within the Orthodox Church created a new interstitial quagmire of theological, legal, and constitutional judgements. The state was keen to downplay and de-politicise the Putin element of the protest so as not to encourage copy-cat protests. In the end the legal line of attack centred on portraying them as rather naïve young women and prosecuting them for causing offence to 'believers' by committing a sacrilegious act in a holy place. To achieve this, the prosecutors had to fabricate a monolithic imaginary of Russian Orthodox believers for whom this act was an act of outrage, and that the performance of the song had also undermined 'the spiritual foundations of the state'. They also had to insert canon (or religious law) into secular law as the means by which to prosecute them. A group of Russian attorneys highlighted this irony following Pussy Riot's indictment:

The investigators accuse these women not of infringing on public order or safety, but of violating the canons of the Orthodox Church. Their behaviour neither contradicted general state order nor undermined public safety. The operation of these prescriptions and proscriptions that they violated extend only to the territory of the Orthodox Church. If they had done the very same thing outside of a church it would not have been possible to accuse them of anything.

(Quoted Uzlaner, p. 50)

Thus, there is a certain paradox in this Russian case study of the postsecular whereby the public sphere of a so-called secular state needs canon law, dating back to the seventh century (the Council of Trullo) to help police its legal borders and provide legitimacy to its prosecutions.

Uzlaner concludes his analysis by suggesting that postsecularisation creates new interpenetrations between the religious and the secular, by which new hybrid spaces and imaginaries are formed. One set of hybridities he labels as 'pro-authority'. Under these conditions, state and religious actors join together to create new legal and cultural narratives whereby the resources of religion and faith are used to support and legitimate the authority and status of the secular state. Another set of hybridities he labels as 'oppositional'. These are new spaces of legal and cultural contestation where the resources of religion and belief combine with other actors to challenge the authority of its state and its claims upon the loyalties and obedience of its citizens to its version of nationhood and national identity.

Religion, welfare, and Russian civil engagement – postsecularity as desecularisation from below

One can observe some similar dynamics of pro-authority and oppositional postsecularity in the ways that faith groups are inserting themselves more fully into the Russian civic sphere via the largescale provision of welfare and other statutory services in the wake of the collapse of the State. This increased involvement has its roots in the 1980s when legislation was introduced easing the restrictions of public religious practice. This lifting of regulations saw the creation of a more competitive religious market with local and global Pentecostal pastors preaching in cinemas and stadiums and gaining conversions. In the late 1990s this influence accelerated with the severe economic and debt crisis that hit the Russian Federation, and that coincided with a health epidemic created by explosion in narcotic use created by the liberalisation of Russia's borders (Atlani *et al.* 2000).

Barbara Giudice identifies three ways in which Pentecostals now find themselves at the centre of Russian and political life; namely social partnership, political partnership, and global networking. These categories highlight the deep and complex ways in which religious practices and narratives are now imbricated in the governance structures of the secular state. Giudice's sample is made up of a number of Pentecostal leaders and local authority managers working in deindustrialised towns in the Urals where social deprivation is high and state infrastructure is in crisis. The infiltration of Pentecostal-run social and health services is high. For example, some estimates suggest that 500–1,000 rehab centres providing support to eight million drug-dependent Russian citizens are run by Pentecostal groups, compared to only three run by central government. (Giudice 2016, 223). Indeed, the dependency on the church to not only run services for the State, what we might call addressing a 'welfare deficit', also extends to state-run public infrastructures such as local authorities and civil service leadership, which struggles to attract people of sufficient calibre and experience. We might call this addressing a 'democratic deficit'. It is not uncommon to find that the head pastor of a Pentecostal church is also acting as a local mayor, or a chair of a public health committee, or as a head of a health and social work department. Pentecostal congregation members are also integral to the efficacy of the social care system in their role as speech therapists or social workers (p. 224).

This interpenetration of the religious and the secular in terms of deeply entwined and imbricated rapprochement means that Pentecostal service providers have learned, like their secular counterparts, to circumvent both legal and illegal (i.e. corrupt) obstacles to their work. For example, despite official prohibitions, much funding from international donors is provided for local social services via local Pentecostal networks with the unofficial blessing of the secular authorities who choose not to ask its origins or provenance. Giudice calls this 'an informal sphere of bargain and favour in order to guarantee the long-term activity of their associations' (p. 227) in a volatile and uncertain public sphere.

There is also a lack of prohibition on proselytising. One pastor of the New Light Church, also director of a state-sponsored children's centre, includes Bible study and a Sunday service on the grounds that this offers a more 'personalised' approach to their care (p. 224). Most agencies in the West would still baulk at such an upfront uniting of 'proselytising with a social work component that is sanctioned by local authorities' (p. 224), but in Russia it is seen as part of the postsecular exchange economy. Having said that, an element of the State's acceptance of proselytisation on the part of the Pentecostal groups derives from the fact that it is 'on message' with the State/Church nexus. As well as highlighting the importance of social responsibility, Pentecostals tend to 'place the family at the centre of the social organisation; their opinions on homosexuality do not differ from those expressed by the ROC hierarchy; they condemn abortion' (p. 229). However, the worsening of international relations between the Russian Federation and the West is likely to mean that greater scrutiny will be applied to the overseas funding arrangements of their work on the pretext of national security, thus jeopardising the long-term future of these postsecular arrangements (p. 233).

6.6 India – discerning a postsecular secularity

There is a growing academic focus on the way in which the postsecular discourse is helping to frame a renewed debate about the distinctive nature and role of the Indian model of secularism. There is an increasing sense that the model that emerged in the post-independence period, influenced by the decisive contributions of both the 'fathers' of the nation, Nehru and Gandhi, has stalled. Having enjoyed a long-term hegemony since Independence through the continued line of the Nehru dynasty, the Indian Congress party steadily lost control and influences during the 1980s and 1990s in the face of a resurgent Hindu nationalism. In 2014, the BJP party, with its explicit doctrine of *Hindutva* (literally Hinduness), swept to power with an absolute majority in Parliament (282 seats to ICP's 44) with the Indian Congress Party more or less wiped out beyond its urban and middle-class strongholds around conurbations such as Delhi and Bangalore. To some extent, the rise of the BJP is not unprecedented. It accurately reflects the current demography of the country, which contains roughly 82 per cent Hindus, 13 per cent Muslims, and 2 per cent each of Christians and Sikhs. The remaining 1 per cent is made up of dozens of minority communities including Buddhists,

Jains, Jews, Zoroastrians, and those professing no religious affiliation. The BJP also traces its roots to the establishment of the Rashtriya Swayamsevak Sangh (RSS), the political movement established in 1925 by Keshav Baliram Hedgewar as a direct response to the threat to Hindu culture presented by British colonial rule. There is therefore a widespread perception of a deep crisis in the Indian project of secularism at its inability to halt the rise of religious fundamentalism. Some hope that a postsecular discourse might be able to refresh and reinvigorate what an Indian secularism might look like for the future.

Historically the Indian idea of secularism has fluctuated between the Nehruvian and Gandhian traditions (Battaglia 2017). Both traditions recognise that a secular constitution has to work with the grain of a plurally religious culture, which had framed a relatively peaceful co-existence before the era of the British Empire. For Nehru, more steeped in the principles of Western secularism, the constitution would be based on a majority/minority principle, namely the upholding of minority rights within a context in which many minorities were in a position of being potentially dominated by a Hindu majority. Secularism within an Indian context therefore does not mean a state where religion is discouraged. Rather, it provides freedom of religion and conscience, including freedom for those who may not have a religion. However, a non-discriminatory state is not the same as a non-interventionist state. In cases where the rights of a minority to freedom of expression and conscience are threatened by the majority, the state in theory will intervene differentially rather than uniformly.

Gandhi, unlike Nehru, was intensely religious and saw religion as an inescapable dimension of all life, including politics. His view of religion was nonsectarian. Rather it was 'a belief in ordered moral government in the universe ... this religion transcends Hinduism, Islam, Christianity, etc. It does not supersede them' (Madan 1997, 235).

Gandhi affirmed the inseparability of religion, understood as morality or *Dharma*, from politics. But at the same time, he did affirm the separability of religion (as denominational faith) and the state. His approach to state-based secularism was strongly influenced by the daily lived reality of communalism (namely inter religious and community violence). For Gandhi, the best way for the state to intervene in communalism was to view the individual citizen as a moral agent. As Madan reflects:

> In Gandhi's judgment, the moral individual was the cornerstone of the good society, provided that he or she was an other-regarding rather than self-oriented individual – a *Satyagrahi* (a truth agent) and not a *sannyasi* (a renouncer). In such a society the state's functions would obviously be more limited.
>
> (2011, 17)

Madan offers a pithy summary of the difference between the two traditions of Indian secularism: 'While Gandhi's view of future society envisaged a religiously diverse political community comprising moral, responsible individuals, the Nehruvian state has moved unfailingly in the direction of a secularised culturally homogenised political community of citizens, defined by fundamental rights' (2011, 19).

Ultimately, Indian secularism is caught between these competing visions of what it fundamentally is. As Amartya Sen has pointed out, the project of Indian secularism contains an intrinsic 'incompleteness' which is both its strength, but also its weakness (Sen 2005). It is now this weakness that is being exploited by nationalist Hindu factions to move India as a whole towards a Hindu rather than a secular identity.

So how is the concept of the postsecular being applied to these existing models of Indian secularism, or as Osuri remarks, how is it being applied in the service of a 'critical secularism' (2012)? The overall aim of critical secularism, she claims, is to break open new theoretical and political ground by 'de-transcendentalising' the secular (see Spivak 2004). This involves recognising both its theological roots within the Judeo-Christian tradition and its use as another form of imposed ideology on India by the West. By de-elevating the secular from its almost religious status, one can critique the way that both religion and the secular have been fetishised by the Indian state, and instead open up a new space for understanding tolerance as a lived and performed practice based on everyday religious practice, rather than an ideological block called 'faith'. This idea of faith as an ideological block was a legacy of British colonialism which conceptualised Hinduism into a Western-style religion (Jaffrelot 1999), and which helped it harden its cultural and ideological identity over and against other identities and practices. This clearly paved the way for the steady rise of the BJP and the Hindutvar ideology since the 1980s, and which has become increasingly premised on the tactics of 'coerced cultural assimilation' of the minority other (Kumar 2008).

But what does tolerance mean in the context of 'postsecular secularism', in which the key obstacle to living together harmoniously is both a fetishisation of ideas of identity, and where a key driver to communalism is the fear of conversion, which then reduces one's power base as one loses 'market share'? There are several ideas. One is to move beyond functional ideas and condescending ideas of tolerance to a place where it is seen as being predicated on a lived co-existence (Kumar 2008). Thus, Nivedita Menon stresses that secularism is ongoing and always in the process of becoming, rather than a teleological project. This involves reclaiming the word communalism from its current pejorative connotations of religious fuelled violence by 'untwisting it from its 'statist and authoritarian discourses' … and reworking it into 'our everyday practices' (2007, 139). But what does this look like on the ground? For Menon, it requires a reformulation of a post-colonial history of the whole of South Asia which eschews the Western obsession with religious identity, the structures of minoritisation, and the idea of the nation state, all of which have toxic postcolonial relationships with ideas of religion and identity. Far better, she suggests, to cultivate a wider sense of ownership of multiple identities rather than a religious one by looking at how Indians also belong to each other regionally, linguistically, and in terms of caste. This would represent an explicit encouragement of hybrid positions where one can own both a secular and culturally religious identity that rejects religious fundamentalism. 'Such an assertion of identity would say I reject the politics of Hindutva as a Hindu and a non-believer, it does not speak to me', or 'I reject the politics of Islamic fundamentalism as a Muslim and a secularist' (p. 137).

From the Gandhian tradition, a secular ethics of co-existence requires a privileging of the non-modern. Two proponents of an emergent 'postsecular secularism' are Ashis Nandy and TN Madan. Nandy refers to the pre-colonial age, when 'thousands of communities living in the subcontinent [learned] to co-survive in reasonable neighbourliness' (Nandy 1997, 162 – quoted Jabir 2015, 150). This is a return to 'an indigenous tradition of religious tolerance' (Jabir, 150). Madan meanwhile returns to Gandhi's concept of a minimalist secular state which he reenvisages as a decentralised polity predicated on a notion of participatory pluralism within civil society (2011, 19). For Madan, Gandhi's secularism transcends political control because at its heart it is a moral value that 'implies reference to a transcendental principle' (p. 18). Critics of this position, however, stress that this is an over-romanticised reading of the past that ignores the inherent hegemony of Brahminic Hindu ideology and its unprogressive legacy before European colonialism (Jabir, 151). It is also interpreted as an ongoing form of majoritarianism, as it leaves intact 'normative notions of tradition and culture that have the emergence of the nation-state as the historical horizon of their emergence and codification' (Osuri 2012, 48).

We can therefore observe that the postsecular debate within India is being deployed as an exercise in 'de-transcendentalising' both the religious, the secular, and the idea of the nation state, weighed down by a heavy religo-secular ontology that is so distinctive to India's postcolonial heritage. Much of the search for a postsecular secularism has been prompted by the steady rise of Hindu nationalism which threatens the Indian secularism project. The de-transcendentalising of the secular has involved an intellectual shift away from 'identity' as something that secularism has to protect and uphold. Instead, it has moved the emphasis towards looking at secularism as 'performativity' in the midst of a progressive vision of communalism. Within this intellectual space, hybrid identities are encouraged and constructive learning of indigenous models of communalism before the advent of colonialism are advocated. For some like Bhargava (2017), echoing more of the Nehruvian tradition, the Indian state's notion of secularism needs to be more confidently rooted in a rights-based approach that addresses the inbuilt discrimination faced by women and minorities, rather than the principled distance model identified earlier. However, many see this as an approach that simply continues to stay in the stuck-binary of majoritarian and minoritarian statuses.

From our postsecularity perspective, there have emerged some tantalising discussions of secularism being first and foremost a form of ethics predicated on ideas of co-existence, and that the way to break out of the current constitutional impasse is to explore practices and tactics of progressive communalism; namely how spaces are actually shared both physically, but also mediated through the mass sharing of religious spectacles on television and social media in an increasingly enchanted mediatised environment (see also Lewis 2016). The ideas of new ethical predispositions and shared spaces as a way of reimaging secularism could be a contribution that postsecularity offers. But much more empirical evidence looking at tactics and subjectivities of hospitality and tolerance that do not rely on a metanarrative of the pre-colonial golden age, and instead looks at the agonistic complexity of Indian society in the here and now would be required. This is so the debate on what an Indian 'postsecular secularism' looks like can be moved away from abstract formulations and more into the realms of the material and the messy.

To this end, Phillipa Williams' (2015) ethnographic research with Muslims in Varanasi, North India is instructive of the everyday relations of receptivity and co-presence that contribute to peaceful relationships. Despite the failure of Indian secularism to protect Muslim minorities from discrimination, Williams' work traces the mundane practices of the state, civil society, and Hindu-Muslim economic exchange in materialising nonviolent resolutions to communal violence. Secularism is mobilised by Muslims as a political strategy, through public protests and less visible tactics and encounters within the local state, to (re)affirm inclusive rights and equal citizenship. Postsecularity, as read through Williams' work, seems to be evident in the informal social ties, the local peace initiatives, subversive tactics, and economic relations that potentially are generative of more hospitable relations to alterity, but also draw attention to the fragility of peaceful sociality.

6.7 Chinese and Hong Kong secularisms – ambivalence, change, and contestation

One of the most illuminating and interesting contributions to emerging understandings of postsecularity at the level of the State is the relationship between religion, belief, and secularism in present-day China. Most commentators agree there are three timeframes relevant to this debate (Chu 2014). The first is the Maoist Cultural Revolution during the 1960s and early 1970s, which saw Marxist critiques of religion as the opiate of the people and the instrument of the ruling class as the pretext for the stripping of powers of religious leaders (including their incarceration, or forcible laicisation). It also led to the public purging rituals and liberating the Chinese people from all kinds of superstition (Chau 2011, 5). Then came the period from the late 1970s, initiated by Deng Xiaoping, which saw market-based economic liberalisation and an opening up of China to the outside world. This approach also sparked a revival in interest in religion. Realising that religious revival was a social reality that was better to work with rather than suppress, the Chinese Communist Party (CCP) rolled out more specific policies on religious observance. This period saw an increase in the number of state-sanctioned religions and more tolerance towards religious activities that were deemed economically beneficial (like the building and renovation of temples, and the practice of religious pilgrimage (Chau, 5)). In other words, the state became 'regulatory and managerial towards religion rather than suppressive and hostile' (Chau, 5; but also see Potter 2003; Li 2006).

The third phase currently unfolding with the election of the current President Xi Jinping in 2013 appears to have witnessed a hardening of the line against religion in the name of nationalism and security. Christian, Muslim, and Buddhist groups have witnessed the demolition of sacred buildings, security clampdowns, cyber monitoring, and bans on overseas travel. Many human rights lawyers involved in high profile disputes with the Chinese Communist Party are Christian.

Within this more hard-line environment, religious belief and practice adopts a number of tactics aimed at 'getting into the official fold' and 'creative dissimulation' (Chau, 6). 'Getting into the fold' involves obtaining a status as one of the five recognised religious groups in China (i.e. Islam, Buddhism, Daoism, Catholicism, Protestantism) and becoming an officially recognised venue for religious activities. 'Creative dissimulation' reflects strategies which are designed to 'disguise one's religious activities as something else that is more palatable in official eyes' (p. 6) – for example, marketing temples or religious festivals with tags that denote 'folklore, museums, charities, tourist destinations and local landmarks' (Chau, 7). Buddhism, which expanded greatly under the Reform era, is currently estimated to have 100 million practitioners (although this official estimation has not changed since the 1990s and so is likely an underestimation). Its revival has been driven by many factors. It has become associated with commercially successful movies in the 1980s linking key monasteries to the practising of martial arts (Zhe 2011, 33–34). These monasteries have now become a global signifier of popular Chinese culture and identity. Leading temples began their own in-house publishing ventures covering such topics as music, medicine, sculpture, painting, printing, and philosophy. These became seen as an adjunct to Chinese cultural studies (Zhe, 39). Buddhist ideas are also seen as synonymous with a growing nationalistic agenda around Chinese culture, sanctioned by the CCP as a 'valuable essence of traditional culture to enrich and develop a new socialist national culture' (Zhe, 40). In the meantime, many Buddhist temples have become sites of popular tourism and pilgrimage, thus further imbricating Buddhism with localised economic strategies much prized by regional CCP leaders (Zhe, 42; Ji 2006). For example, China's Buddhist Association and the Chinese central state played an important geopolitical role in supporting the building of the Tian Tan Buddha (1993) in Po Lin Monastery, Hong Kong – representing a convergence of interests between the desire for closer social and cultural ties between mainland China and what was at the time colonial Hong Kong (Qian and Kong 2018). Anxieties surrounding global tourism in the early 2000s led to the Hong Kong government adding a cable car and tourist village adjacent to the Monastery to capitalise on 'authentic' Buddhist culture. More recently, the Hong Kong Government, with backing from the Chinese central state, granted the development of a new five storey cultural and religious centre including a 'Ten Thousand Buddha Hall', which contains 10,000 gold-gilded statues of Buddha and precious Buddhist relics and arts, at the cost of 400 million HKD. This temple project is thus imbricated in spectacle and heritage tourism, as well as land development and pricing processes. Qian and Kong frame the debate, not as a 'secular vs. religious hermeneutic', but as the collision between two competing, yet similarly entrepreneurial logics – the developmentalist logic of the state and the monastery's own vision of expanding and consolidating a 'spiritual marketplace' (161).

By reinventing itself as an innate part of Chinese traditional culture, Buddhism has reproduced itself across the nation, including at the highest intellectual and cultural levels. It has forged a productive and largely harmonious relationship with the Chinese political and cultural elite. However, this cultural appropriation of Buddhism also risks minimising or relegating its transcendent, mystical, and ethical side, thus producing a 'superficial' Buddhism (Zhe, 46). Cultural capital is converted into economic capital by the relentless marketisation of Buddhism for jobs and tourism. Buddhist symbols have been de-contextualised from their sacred settings to become ubiquitous within secular discourses. Zhe concludes thus: 'Buddhism has revived in contemporary Chinese society, and even now thrives, but at the cost of eroding its own religious foundation' (p. 47).

Somewhat different trajectories of postsecularity are offered in the case of the official (as opposed to the underground) Catholic Church in China. Longitudinal ethnographic research into Catholic communities in Shanxi province in North China shows how it has embraced narratives of modernisation from both Vatican II and the Chinese state. Vatican II's exhortation towards 'modernisation' has encouraged the church to build new sacred spaces that combine clear elements of Chinese cultural design in their architecture, deploy Chinese as the language of the Mass, and introduce the Virgin Mary as the centre of worship rather than Jesus Christ (Harrison 2011, 206). Other features of official Chinese Catholicism, which sympathetically mesh with wider cultural norms as sanctioned by the CCP, include the introduction of a grave-sweeping ceremony to coincide with the Christian feast of All Souls and the publication of Catholic lineage genealogies going back to the conversion of a family member, and which in turn go back to Adam and Eve (Harrison 2011, 213). The tracing of genealogies is a precise echo of practices and privileges associated with the Chinese elite in society (214). In this way, Chinese Catholics intentionally assert a hybrid identity that combines elements of modern and Chinese identity, whilst also affirming loyalty and belonging to the transnational Catholic community.

Harrison concludes that the emergence of the official Catholic Church reflects 'a dynamic interaction between traditionalism and reformism' (p. 218). However, this positive view is countered by other scholars in the field (e.g. Madsen 1998) who interpret the Catholic Church's conservative and hierarchical structures and traditions as deeply inimical to the development of civil society in those parts of China where it is dominant. As more recent scholarship suggests, the relationship between the Chinese Catholic Church and the Chinese state is complex and multi-layered, and incorporates postsecularist dynamics of both accommodation and confrontation to state-sponsored secularism (Chu 2014, 20).

A further spatially-inflected case study highlighting how state narratives of secularisation and secularism are being renegotiated by practices of local postsecularity comes from Gao, Qian and Yuan (2018) who explore the impact of rural migration to a newly industrialised village as part of the economic expansion of the Shenzen Special Economic Zone. The village in question, Shanzuli, was, prior to the Cultural Revolution, a showcase village for Christian missionary activity. The interviews and case studies of newly arrived migrant workers, and their reinvigoration of the local church is located within a complex set of intersections within the village; between nostalgia, and a modern-day *ennui*, particularly on the part of the newly affluent middle class. The authors locate this local to-ing and fro-ing between religious and secular narratives of identity within an overall political metanarrative of different secularities. With reference to subaltern agency and autonomy expressed by the Christian converts, they suggest

> that secularisation is a multi-layered social process that has different manifestations and ramifications at different scales … it is not a homogeneous process which flattens out spaces of values and ethics: in Sanzhuli, while native villagers find comfort and new anchors of identities in material affluence, migrant workers and some young villagers revive religiosity to engage with the existential conditions of industrial capitalism.
>
> (2018, 14)

Alongside highlighting the ways exploitative labour practices are theologically legitimatised by some migrant workers, the findings here reveal distinct opportunities for emergent postsecularity to cut against capitalist and 'otherworldly' religious enchantment and instead embody new forms of sacralisation – in this case, of the migrant worker – that challenge state-propagated second-class treatment of migrant workers (Gao *et al*. 2018).

These Chinese examples of state-sanctioned secularism suggest that despite the government's best efforts to control access to religion and belief, or to co-opt it towards supporting the ideologies of the state, interest in and affiliation to both official and unofficial 'religion' in China is thriving in terms of individual popularity. At present, the growing revival in Confucianism, and to a lesser extent Daoism, appears to be a state-sanctioned attempt to reinforce values of loyalty to the traditions and expectations of family elders and social norms (Hammond and Richey 2015). The question not yet able to be answered is the extent to which a Chinese version of postsecularity (namely an official toleration of religion and belief within a normative Marxist ideological framework alongside the strong growth in populist religious and spiritual re-enchantment) will also enable new forms of political and cultural agency in wider society. Much depends, one suspects, on the extent Chinese citizens and, society as a whole, can evolve a rapprochement between religious/secular identities in which it is possible to be both a loyal member of the Chinese secular state and an open follower of a religious tradition, and where affiliation to that religious tradition is seen as a civic asset, rather than an obstacle to public participation.

Unlike mainland China, the possibilities for postsecularity in the Special Administrative Region of Hong Kong are deeply shaped by the state-religious nexus of the British colonial period. Under British administration, Christianity was considered the 'default' religious belief for elites in colonial society, and the Hong Kong government restricted the building or expansion of Chinese temples resulting in most new urban areas not having neighbourhood temples. Alongside rapid industrialisation which further disrupted Chinese folk religious practices in Hong Kong, the anti-traditionalism and iconoclasm of the Chinese Communist Party brought a 'modernising' impulse to mainland China whereby religion was considered superstition. Mass migration from the mainland to Hong Kong in the 1950s resulted in a complex secularised and multi-religious urban landscape – encompassing from iconic Buddhist and Daoist temples, Confucian ritual traditions, Catholic Cathedrals and evangelical megachurches, 'informal' places of worship nestled down busy streets, alongside 'everyday' religious customs such as burning joss paper and making material offerings to deities to ensure good luck and fortune. Hong Kong is a global and multicultural city animated by immigrant religious subjectivities, most clearly exemplified by Filipino domestic workers whose religious belief and communal ties are active in home and place-making strategies (Lindio-McGovern 2004). More recently, the transitional arrangement of 'one country, two systems' has meant religious practices that were once denigrated and overlooked in Hong Kong are now becoming important demarcations of cultural heritage – serving as key symbols of Hong Kong identity as distinct from mainland China.

The hybrid religious and spiritual landscape in Hong Kong provides a rich terrain for the emergence of postsecularity, which, can be most clearly demonstrated in the Umbrella Movement in 2014. This series of protests was driven by secular pro-democratic aims, most notably, for Beijing to grant universal suffrage to residents of Hong Kong. Although only an estimated 10 per cent of Hong Kong population identify as Catholic and Protestant, several scholars have traced the key role Christianity played in the Umbrella Movement with regard to leaders, political rhetoric, and spiritual activism that sustained protest beyond institutional religious support (see Tse and Tan 2016; Ng and Fulda 2017). Political identifications of religious groups in Hong Kong tend to portray Christianity as pro-democratic whereas Buddhist and Daoist groups are seen as more pro-establishment (e.g. Hong Kong Buddhist Association and the Po Lin Monastery have been vocally supportive of Hong Kong government in calling on the initial Occupy Central protesters in 2013 to back down; see Qian and Kong 2018). It is important to note, however, the significant divergences among Christian churches and leaders in relation to the Umbrella Movement (Chan 2017) and recognise that on the ground, the protest was animated by individuals from all faiths and none, including those in disagreement with the 'official' line of the respective Buddhist and Christian traditions and denominations. The Umbrella Movement manifests numerous traces of emergent postsecularity. Tse and Tan (2016), for example, trace the theological significance of the Hong Kong Umbrella Movement, its broad-based alliances across lines of social and religious difference (Tse 2015) and the 'crossing over' of religious and secular symbolism, people, spaces, and narratives. In an act that blended parody and sincerity, for example, protestors erected a small shrine to Guan Gong – a warrior deity representing justice and benevolence, and commonly worshipped by the police – thus representing an iconic challenge to police brutality and a symbolic reversal of power: '[t]he police's god of protection has left them and joined the protestors, for justice is on the side of the protestors' (Kung 2016, 112).

6.8 Postsecularity and the religious reterritorialisation of public space

Our final element of analysis is to consider how conditions for postsecularity are shaped by the spatial and sensory dimensions of religious reterritorialisation of public space. In recent years there has been growing awareness of the material and affective impacts of religious practices and belief on the urban environment; namely how the 'religious' produces the 'urban' and the 'urban' produces the 'religious' (Garbin and Strhan 2017). The nexus of literatures exploring this spans sociology, religious studies, critical and human geography, cultural and post-colonial studies, and anthropology (Cloke and Williams 2018). We are interested in how material and sensory aspects of the 'urban' shape capacities for receptivity, rapprochement, and (re)enchantment.

In a chapter for a collection of essays and artistic enterprises exploring these issues (Global Prayers 2013), Birgit Meyer echoes many of the working assumptions in our volume; namely that the concept of the post-secular as originally deployed by Jürgen Habermas creates an intellectual and conceptual caesura in the uncritical and hegemonic discourse associated with the secularisation thesis. Habermas' ubiquitous definition refers to the need to 're-imagine' the public sphere in which the 'vigorous continuation of religion' (and we would add belief) within a 'continuously secularising environment must be reckoned with' (2005, 26).

Meyer agrees that Habermas' category of the postsecular, 'acknowledges the presence, force and potentially positive influence of religion in the public domain' (2013, 592). However, rethinking the role and place of religion 'as a public force' needs to move beyond mere acknowledgement into a new framework that critically studies religion and belief as an 'historically and culturally situated phenomenon'. This involves understanding how 'religion' and the 'city' are 'interrelated in various settings' (p. 595). This newly understood configuration unearths two major research questions, namely: 'How do new religions transform urban space? And conversely, how do 'cities generate specific forms of religion?' (p. 595).

In addressing these questions, the Global Prayers volume expands the range of globalised spaces where the interface between religious and secular practices, backdrops, and actors can be researched and theorised. New sites of investigation include Berlin, Lagos, Istanbul, Beirut, Atlanta, London, Tehran, Kinshasa, Jakarta, Mexico City, Mumbai, and Rio de Janeiro. Out of these many case studies, two in particular offer contrasting understandings of material postsecularity.

The first is an ethnography by Anne Huffschmid exploring the rise of the 'secular' saint Santa Muerte – or Saint Death – in some urban conurbations in Latin America. The practice of creating spaces for informal public worship of a 'secular' saint, including the deployment of processions and pavement statues, rosary beads, and prayer circles is not so much replacing, as complementing, the traditional veneration of the Virgin of Guadeloupe in cities such as Mexico City and Buenos Aires. Huffschmid locates her enquiry within an astute understanding of the role of liberation theology in Latin and South America. Liberation theology enabled the Catholic church to articulate and harness the aspiration of the urban poor via what came to be termed 'base communities'. This missional strategy was particularly effective in areas ravaged by drugs, precarious employment, and impending demolition or gentrification. However, liberation theology concepts of community emancipation, which borrowed heavily from Marxist materialist and class analysis, have now become undermined by increased religious competition (particularly Pentecostalism) as well as more individualised and psychological interpretations of liberation (2013, 397). Some anthropologists have blamed the decline in liberation theology's appeal precisely because its Marxist materialist analysis has de-emphasised the traditional appeal of 'milagros' or miracles and has 'neglected religious emotions and the need for spiritual experiences' (p. 399).

Within this context, populist veneration of Santa Muerte has emerged out of the close cultural association with All Saints Day. The veneration of Santa Muerte on All Saints Day at newly-created shrines incorporates many elements of traditional Catholic veneration:

> the Rosary and Pater Noster, the sign of the Cross and blessings, the gesture of moistening the mouths of the statues with 'holy water' ... and the form of the religious procession [as] pilgrims slide towards the saint on their knees, at least for the last few meters.
>
> (p. 404)

There are several official Catholic denunciations of the growth of the Santa Muerte cult on the grounds that it represents spiritual decline, a loss of values, and generally perceived 'diabolic' competition (p. 399). However, Huffschmid suggests that the sacred hierarchy of the Catholic faith remains intact, in that those who adhere to the veneration of Santa Muerte do not elevate her above Jesus. He 'remains the highest authority, with the Santa Muerte being nothing but a mediator' (p. 404). The Virgen de Guadalupe is still responsible for miracles, whilst the Santa Muerte is 'more concerned with helping people cope practically with life: success at work, protection from illness, accidents, attacks' (p. 405). Huffschmid concludes that Santa Muerte provides,

> ... an almost therapeutic function for those traumatised both by acute crises ... and the latent state of crisis that Mexico is always in owing to the

obscene gap between prosperity and extreme poverty, impunity and corruption.

(p. 405)

The second case study focusses on the ethnographic work of Ayşe Çavdar, a Turkish and self-identified secular housing activist, and her efforts to get inside, both physically, but also emotionally, the modern religious gated communities for the aspiring Muslim middle-class. Başakşehir is the first largescale housing project of the Islamic RP or Welfare Party. Çavdar wanted to gain access to the project (on the outskirts of Istanbul) to find out 'what does Political Islam mean to city dwellers and in particular, what does it suggest to those who are devout' (p. 200)? The development has been the centre of vociferous debate between secular and religious political actors in Turkey, with those on the former side branding it a 'no-go' area. The particular area of Başakşehir that she wanted to explore was a housing development (Oyakkent) that was funded by a pension fund for retired military officers and soldiers, many of whom saw themselves as upholding the virtues of the secular vison of modern Turkey. On the basis that the wider Başakşehir development was becoming more self-consciously Islamic, the officers decided that they did not want to retire there, so instead sought to protect their investment by renting the properties out. Within their tenancy agreements they attempted to stipulate that no tenants should seek to wear or adopt overt religious practices or dress, including the headscarf. The families moving in had to be 'modern' and to have only a few children (p. 203). This tactic failed, and the 200 or so 'secular' families unable to move because of the shifts in the housing market had to stay, seeing themselves as a small minority within a religious majority. As Çavdar explains:

> In short Oyakkent turned into a battleground of lifestyles, between the rep-resentatives of an institution that saw itself as the guardian of secularism and representatives of religious piety, a group that was fast becoming a middle-class majority.

(p. 204)

However, Çavdar herself often met with responses of indifference and hostility whenever she tried to engage in conversations or join the very few public spaces that revolved around family, and women in particular (namely play parks). The men went to work early in the morning. The communal facilities were rarely used as people did not feel comfortable with being potentially observed in public spaces. Çavdar describes her attempts to join a women-only Quran reading group which a friend offers to convene in order for her to get access to some interviewees:

> Seven women ... met at my neighbour's flat ... my neighbour cooked for us all and some of the other women brought dishes they had made at home. We read the Quran together for half an hour and then we removed our veils and

started to talk ... I introduced myself and told them what I ... did professionally, only later to explain about my research. Two women in particular, both of South-eastern Anatolian origin, were not happy to hear that there was a researcher at the gathering; they remained silent for the rest of the meeting. The others began to tell me about their experiences of Basaksehir: nobody knows each other, no one intends getting to know each other; public transportation is very bad. ... I then felt that I needed to ask why they had come here; it was quiet, peaceful, planned, secure and a good place to raise children were among the reasons given.

(p. 205)

It is important to remember that many of those who moved to this middleclass enclave were themselves original residents of the *gecekondu* (or shanty towns); working-class communities, who immigrated from the rural areas to help build Ataturk's vision of a modern cosmopolitan and urban Turkey (p. 199). It was from these very areas that a newly politicised Islamic identity emerges that will eventually propel the current longstanding President of Turkey, Recep Erdogan to power.

Çavdar writes:

...I wanted to find an answer concerning the dynamics that had brought about political Islam and what kind of transformation it had undergone since the 1990s. My interviews with the pious and middleclass of Basaksehir [showed] it was obvious that they were not keen on the perception that they were poor and excluded from the city. They had come ... to enjoy their recently elevated position, won through collective political struggle and hard work, which had boosted the national economy while also improving the local economic situation. They did not want their lifestyle judged by the secular lobby or by those with traditional understandings of piety, who looked down on the conspicuous exhibiting of newly acquired social positions and power ... they wanted respect, not only for their religious and ideological backgrounds but also because of their success commercially. A woman explained to me how it is possible to merge religious and profane values into one value system. 'Allah en iyi nimetlerini ona inanananlarin üzerlerinde görmek ister' ('God wants to see his best blessing in the believers.').

(p. 207)

In this case study we observe how different fault lines come together to create spaces of postsecular negotiation; class, gender, ideology, and planning style intermingle to create new ethical subjectivities (modern public piety) and new types of space (the majority-religious gated community). In this case, these new subjectivities and spaces appear, not to produce an excess of progressive social capital, but are instead highlighting and exacerbating existing social and cultural tensions within the wider society.

Another vista of material urban postsecularity is opened up by Yamini Naray-anan's edited volume *Religion and Urbanism – Reconceptualising Sustainable Cities for South Asia* (2016) which explores the relationships between religion and urban development in South Asian cultures. The way that Narayanan defines the relationship between religion and urbanisation is primarily along the lines of heritage and informality. She makes a connection between religion and spaces of informality which she calls 'grey zones'. These are both spaces of resistance, but also spaces of exception that collude with the oppression of animals and women (p. 144). Her call is for planning to develop a theological as well as anthropolog-ical understanding of the role of religion and belief in the way in which it shapes identity and space. Postsecularity in this context implies that religion, modernity, and post colonialism are powerful forces that are shaping the new megacities of the world. The production of new global megacities is precisely coterminous with those parts of the world witnessing the highest rises of religious affiliation through population growth and other globalised processes. Her starting point is that religion in South Asian cities infiltrates every aspect of planning and urban development and she deploys a typology developed by Michael Pacione (1999) to flesh this idea out in her volume. Pacione looks at the impact of religion on the built environment in terms of both *tangible* and *intangible* heritage. Tangible heritage pertains to the cultural and economic impacts of physical buildings and practices (such as pilgrimages and other forms of public ritual). These spaces, which were often contested sites of resistance during phases of colonialism, now become new sites and sources of post-colonial nationalism, for example, in cities such as Colombo and Yangon. Intangible cultural heritage refers to the many ways that material impacts associated with 'ritual' and 'social institutions' are mixed with non-material sources for some of these impacts; 'doctrine', 'sacred narrative', 'ethics', and 'experience'. This is especially true in cities such as Bhakatapur, Jaipur, and Amritsar. For example, with respect to Bhakatapur, urban anthropologist, Steven Parish, observes how these tangible and intangible features combine to reflect an ontologically charged space in which secular activities are contained and interpreted in order to give a coherent narrative of the heritage of the city. The Hindu pantheon of gods and goddesses, and the rituals, actions, values, and belief systems embodied on a daily basis in its built form engage with the topography of the city. They are 'material witnesses to a variety of historical projects and happenings, responses to and indexes of the economic, utilitarian and communicative needs of the city' (1997, 453 quoted Narayanan, p. 13).

This section has demonstrated how religion 'produces the urban', and reterrito-rialises public space on different scalar levels, and within different urban publics (the Mexican downtown, the Turkish exurb, the south Asian pilgrimage city). We can observe how the Santa Muerte public rituals reflect a performative space in-common, a 'thirdspace' that represents the crossing over of religious and secular/ working class tropes into the creation of hybrid subjectivities that are associated primarily with the therapeutic – namely healing and cleansing. The Başkaşehir case study highlights the obstacles, rather than the encouragements towards a postsecular rapprochement – and how some urban planning strategies create invisible barriers to the formation of trust and interaction (for example, a lack of shared public space, or the lack of a coherent shared narrative). In South Asian cities this notion of a shared and symbolic space is mediated and upheld by a rich public and symbolic art.

6.9 Conclusion

This chapter has emphasised the importance of attending to the wider political and social conditions which shape the capacities for emergent postsecularity in different settings. It is essential to grasp the varieties, and uneven manifestations, of secularism and secularity *between* and *within* different states, and resist any normative application of postsecularity that is insensitive to place, identity, and culture. While the co-constitutive dynamics between religious identities and practices, and modes of secularism and secularity are not new, we suggest the above case-studies suggest the differentiated and lived out fields of the religious and the secular identified by Bourdieu, Casanova, and Taylor, and so characteristic of previous understandings of modernity or modernities, are becoming increasingly blurred and hybridised across global settings. This creates a highly diverse yet ambivalent terrain for subjectivities of postsecularity to emerge, where capacities for rapprochement, receptivity to difference, and hopeful enchantment are locally configured and circumscribed by changing social, cultural, and political-economic geographies. At this point, it will be instructive to turn to a source that explicitly attempts to work with ideas of the postsecular and the postcolonial. In *The Postsecular Imagination – Postcolonialism, Religion and Literature* (2013), Manav Ratti attempts to develop and define what he calls a 'secular postsecular imagination'. His literary project engages with major religious perspectives and identities such as Buddhist, Sikh, Christian, Muslim, and Hindu. These debates are in turn brought to life in novels in which cities such as Shimla in India, the Golden Temple in Amritsar, and the Ta Prohm Temple at Angkor in Cambodia play a defining role.

For Ratti, the 'post' in 'postsecular' can be defined in the following interlocking and multifarious ways. As already outlined in Chapter 1, he argues that the 'post' does not represent the teleological end of secularism or its replacement by religion. Neither is it a return to an uncritical Western concept of Enlightenment that imagines the secular 'as the sole bearer of rational progress' (p. 21). Rather Ratti sees 'postsecular' as a term of negotiation; namely religion's negotiated relation with the secular in the form of what he calls ''new combinations', [emerging from] 'social crisis' and '[interfusing] the other'' (p. 22). This idea of negotiation helps prevent the deconstruction of the secular as the received opposition of the religious and vice versa. For Ratti, postsecularism is one of the 'destinies' of postcolonialism. He suggests that if literature creates a forum in which

... new conceptions of secularism and religion can emerge, then new forms of 'ethics' will also grow that not only reject the national violence of the past done in the name of either one or the other, but also nurture 'the hope for a better future for all'.

(p. 7)

We have nuanced Ratti's rather abstract call for a new post-colonial postsecular imagination with five lived dimensions of postsecularity derived from our exploration of non-Western, and where possible, non-Judeo-Christian debates concerning the postsecular. First is the strong sense of religion being perceived as a threat to established secular orders in nation states in terms of national security and radicalisation. However, the more the State attempts to control religion and belief, the more it simply re-enters the public sphere in adapted ways, such as cultural heritage or tourism, welfare engagement or political art. Second, is the appeal to alternative ethical subjectivities and understandings of agency that the postsecular turn demands. This is particularly true of the ways in which non-Western feminist notions of power, agency, and autonomy, along with ideas of intersectionality are being reformulated. Third, is the emphasis on religious/ secular hybridities, which are by and large absent from Western discourse and which highlights the postcolonial dimension of the discourse. Fourth is the growing sense that the idea of the nation state also needs to be reimagined if a new way of dealing with diversity is to be achieved. For example, we have seen how eliciting notions of lived accommodation within localised performances of tolerance, hospitality, and the sharing of space are being increasingly scaled up to offer policy clues at the macro level. Fifth, it encourages a focus on the inter-penetration of the religious and the secular at the level of material production of urban and social space, thus reinforcing our belief that postsecularity is a way of being that expresses itself in new material structures, ethical practices, and forms of governance. We now take some of these themes into our final chapter, where we critically define the emerging contours of postsecularity as the basis for a sustained and credible politics of hope.

Note

1 This conference was organised by the Institute for Religion, Culture and Public Life, University of Columbia, 10/11 November 2011.

7 Conclusion

7.1 Introduction

In this book, we have developed the concept of postsecularity as a framework through which to render visible and interpret ethical and political subjectivities emerging between boundaries of religion and secularity. Our approach contrasts with popular interpretations that claim the postsecular as an empirical description of reality or epoch shift that refers to the renewed visibility and presence of religious voices in late secularised society. Rather, postsecularity refers to a way of being characterised by the acceptance and living out of receptive generosity, the capacity for (if necessary sacrificial) rapprochement, and the commitment to a re-enchanting reshaping of desire away from the values of self-interested capitalism towards counter-cultural values of hospitality, generosity, and justice. Postsecularity, we have argued, is most visible in actual embodied tactics, affective registers, ethical dispositions, and material practices in which faith and secular actors relate differently to alterity and enter into a re-territorialisation of the comfort zones of belief, identity, and ethical praxis.

We have argued that the capacities for, and the political relevance of, postsecularity have been heightened because of the unique moment that we are in culturally and historically, especially in the Western world. Culturally, postmodern curiosity directed at a more hybrid, connected world, and despair at neoliberal hegemony and austerity, have combined to generate more widespread attention on spirituality and religion as sources of hope for political activism. Historically, the dialectical relationship between religiosity or spirituality as a way of being, and secularism as a system of political organisation is in a stage of significant potential renegotiation. After long periods in which religion has been characterised for some by its duress on people's freedoms and its capacity for grand-scale antagonism, its public remit has been usefully curtailed by secularism in order to accommodate greater pluralism in society. However, in its truncated form, significant elements of religion have had time to reflect upon its keenest insights and speak back to secularism, highlighting in particular its overinflated claims to rational supremacy (Connolly 1999). Hence, the relation between secularism and religion has come to a point where at least some of those who are invested in the structures and practice of secularism now seem willing to reconsider the richness of religion in its appreciation of mystery, its capacity for ontological humility amongst the interdependence of all things, and its ethical reflexivity. Moving forward, it is crucial to think of new ways in which religion can be accommodated within secularity that are critically appraised and comfortable with ethical plurality (Hauerwas and Coles 2008). In this volume we have attempted to shift understanding of the postsecular from a relatively fixed conceptual framework of religious-secular relations, towards a more complex ethical role. This role facilitates a renegotiation at various levels of society which re-evaluates not only the relation between religion and secularism, but also the attendant tensions that the religion-secularism relation convenes between ethics and politics, faith and reason, and mystery and structure.

Therefore, in this concluding chapter we want to direct our argument at two strategic targets. In short, we strongly suggest that the spaces and networks of materiality, affect, and imagination that we have identified as postsecularity – and the practices that emerge from them – are both making a significant contribution to the contemporary political juncture, and generating possibilities for enduring political applicability. Through two concluding exercises, we want to substantiate these two points by asking the question: how can the three modalities of postsecularity that we have used as guiding themes in this book – re-enchantment, receptive generosity, and rapprochement – be put to use? First, we will briefly review the arguments that we have made so far in this book; not as an unproductive reiteration of what has already been said, but as a more precise evaluation of the political and ethical salience of each chapter. This reflection will focus in particular on the ways in which practices of postsecularity are currently mediating the various tensions in the religion-secularism relationship. Second, we will identify four ways in which postsecularity can be tactically deployed in political and academic geographical work with durable consequences, suggesting that it has a number of ontological principles which resonate with action research, and therefore provides a territory in which to blend the political and the ethical with the geographical.

7.2 Postsecularity: a hopeful and non-binary ethico-politics

In this section we focus on the ways in which each chapter of this book has built on the last to generate a sense that postsecularity is a hopeful and non-binary way-of-being that troubles neat divisions between religion and secularism, ethics and politics, faith and reason, and mystery and structure. We want to highlight how our discussions in each chapter illustrate how postsecularity, through its generous brand of ethico-politics, is contributing to the contemporary political juncture through its renegotiation of the interface between religion and secularism.

In Chapter 1, we highlighted the political impact that postsecularity has in terms of its analysis of the current political conjuncture. We began by examining the interconnections between religion, secularism, and public action (particularly in care, politics, and charity) in order to highlight the political and theoretical themes that we wanted to affect and progress. The first of these narratives suggests that religion and secularism have a long history of co-producing each other, thus rendering as fallacious any claims that such co-production *per se* is a novel phenomenon. Religious actors have had a longstanding presence in the public spheres of care, politics, and charity and just because there is renewed academic attention upon these relations does not mean that analysts need the new label of 'postsecularity' to describe them. Second, co-productions between religious and other political actors are a risky business. As well as some notable contributions to progressive politics – for example, in the civil rights movement and anti-war campaigns – religion has intensified illiberality through theocratic discourse and heavy-handed moralising in phenomena such as America's religious right. However, we suggested that despite this mix of political outcomes, postsecularity attracts our attention because of its potential to play a progressive role in the current political and cultural moment. At present, the political left and right both offer unappealing affective conditions. The left often presents an ideologically rigid, paranoid factionalism despite being driven by justifiable despair at plummeting living standards, the growing influence and unaccountability of the super-rich, and ecological ruination. The right typically espouses commodity fetishism, inhumane metrics for measuring self-worth, divisive and alienating individualism, and a spitefully defensive belief in the economic and cultural status quo. We argued in Chapter 1 that even though religious involvement in the public sphere can produce problematic phenomena, the current political conjuncture – marked by neoliberal instability – has led to various cultural and academic actors reaching back out to religious organisations and ideas for inspiration, recognising both religion's co-productive relationship with secularism *and* how this relationship might usefully be renegotiated for progressive political ends.

These realisations have led to more widespread recognition that the secularisation thesis – in its classic Weberian formulation – is not to be thrown out but is not beyond questioning, and moreover is capable of challenging whether a fully disenchanted society is possible or desirable. Not only does this political conjuncture foreground postsecular renegotiation of the secular/religious interface, but moreover, subjects steeped in postmodern cultural power-relations are increasingly questioning the metanarrative of rationalist political and ontological authority claimed by secularism, thus accelerating the hybridisation of secular citizenship and spiritual praxis. We posited that postsecularity emphasises this renegotiation and hybridity between religion and secularism, suggesting that its progressive potential lies in facilitating new cocktails of faith and reason, mystery and structure that might provide trajectories past the defensive anxiety of the left and divisive selfishness of the right into non-superstitious forms of enchantment that reinvigorate ethically attuned regimes of desire. We identified three patterns of linkage between postsecular thinking and nascent postsecular empirics – receptive generosity, rapprochement, and prepolitical assemblages – that indicate political ways of being with an ethical anchor. These three modes of action associated with postsecularity allowed us, in the rest of the book, to explore the ways in which postsecularity transcends binaries between religion and secularism, mystery and structure, faith and reason, and ethics and politics. We argued that using these three currents to analyse the deconstruction of these binaries illustrates how postsecularity both *exercises* and *fosters* hope as those practicing it forge an ethico-politics that seeks to shatter the affective malaise of contemporary neoliberalism.

In Chapter 2, we reviewed some of the key theoretical debates around postsecularity in order to further clarify what we perceive to be its progressive political and ethical potential. We first juxtaposed Rawlsian and Habermasian notions of democracy and the role religion plays within them. Rawls' concern is that religion and complex philosophical or metaphysical concepts have little substantive moral content and are difficult for most people to understand. Therefore, he concludes that these ways of thinking are not useful for generating broad understandings of common life, ruining the possibility of political community, dialogue, and democracy. However, Habermas suggests that it is impossible to disentangle the public sphere from metaphysical ideas – they are ubiquitous – and that what is important is to *improve* how secular structures interact with religion rather than unequivocally diminishing this relation. Furthermore, he argues that the metaphysical, ethical, and moral contents of religion and other philosophical positions *can* co-produce crossover narratives (around, for example, justice and dignity) that generate enthusiasm for democratic engagement and civic-mindedness. Through reflexive dialogue about these metaphysical positions, common prepolitical proclivities can be drawn out that move beyond acquiescence to any kind of a/religious fundamentalism. However, this negotiation is full of tensions and we acknowledged the difficulty of working through discussions about the public role and voice of religion; nuance and close analysis is paramount.

We demonstrated this in our second line of argument which explored the purportedly progressive political theology of radical orthodoxy. Although radical orthodoxy postures as a useful way of archeologically reclaiming the Christian values of Western society to redeem civic life, we argued that it is not clear how its intellectual frame might translate into practices that engage contemporary civic plurality without a problematic proselytic edge. As a counterpoint to radical orthodoxy, we examined the 'weak theology' of Caputo and Vattimo that promotes a praxis of faith-in-love that embraces ethical engagement amidst uncertainty and mystery rather than a practice governed by a series of biblically-derived moral tick-boxes that may seem anachronistic or illiberal to many people. Through our comparison of radical orthodoxy and weak theology, we illustrated how different formulations of Christianity might contribute to a postsecular debate about common values and that close readings are necessary to discern what is helpful. We argue that although radical orthodoxy exemplifies a postsecular tendency in its attempt to renegotiate the relation between secularism and religion, perhaps weak theology better represents a postsecular ethos in its more generous posture towards uncertainty and collectively discerning the contents of civic values. We consider that this debate clearly illustrates the 'messy middles' that postsecularity plunges subjects into, exemplified in particular by the new prominence of faith groups in the provision of care in response to an increasingly rolled-back, austere state. In negotiating between Christian religion and secularism, very different outcomes can be produced. On the one hand, we can envisage a 'heavy' postsecularity of theocratic claims to moral authority (for example, much radical orthodoxy), and on the other a 'light' postsecularity that is open to the *performativity* of engaging with difference and has faith in the ability of this process to co-produce something novel and mutually enriching (like weak theology). We argued that 'light' postsecularity goes deeper than 'heavy' postsecularity into the renegotiation of the interface between secularism/ religion. It moves past historical connections between secularism and religion that try to reanimate archaic frames for societal order, and towards a mutual performativity that initiates and deepens ethical proximity. This 'light' postsecularity, we argue, is the basis for a non-binary politics that can usefully renegotiate secularity and religion in the midst of messy middles.

In Chapter 3, we explored the capacity of postsecularity for generating counter-neoliberal *subjectivities*. We began by highlighting the deleterious effects of neoliberalism on the existential and phenomenological life of religious and secular subjectivities. Neoliberalism must not only be understood on the level of policy, ideology or governmentality, but also as an affective politics that works on the register of the aesthetic, ethical, and spiritual. Affective life shaped by the regimes of desire and immaterial labour under neoliberal capitalism does not necessarily produce singular subjectivities but instead, we suggest, has generated a series of ethical capacities characterised by demoralisation, ennui, and ressentiment. Ethical relations of self-to-self and self-to-other are circumscribed and mediated through evermore individualistic and suspicious metrics of deservingness, which flow into practices of self-aggrandisement and charitable sentimentality. The affective politics of neoliberalism therefore is a fundamentally spiritual concern as it seeks and struggles to sever the connections between individual desire and collective life. Rather than championing the ways desire can produce dispositions of being in-common and foregrounding the emergent synergies of this relation which endow a person with purposeful energy and meaning, neoliberalism presents a tired selection of progressively ineffective and instantly-gratifying dopamine buttons that detach people from life in-common through distraction, denial, and ennui. Neoliberalism animates these shallow reservoirs of false enchantment, deflecting attention from the structural causes that underpin the grinding conditions which its austere politics are imposing on an expanding precariat. We argued that these frantic-yet-boring, iterative neoliberal mechanisms are being fractured by a groundswell of citizens becoming disenchanted with neoliberal excess and the stacking up of injustice, that inspire a phenomenological response to 'do something'. As entrenched austerity and precarity affect greater numbers of people, the collective grief that emerges in response to an overbearing and unjust regime results in a burgeoning phenomenology of need. More people are joined by a common quality of suffering; of feeling limited, hemmed-in, desperate, and helpless under neoliberal governance. Postsecularity is not the only response to this affective condition, and we emphasised that existential ressentiment, when combined deep-seated racism, fuels fascist sensibilities and the contemporary rise of authoritarianism. We suggested postsecularity – as one trajectory emerging in this context – presents four possibilities. First, the affective condition of austerity and its attendant material inequalities has the capacity for increasingly diverse groups of religious and secular minded individuals and organisations to enter rapprochement, which in turn can serve as a catalyst for new theological and secular configurations and practices. Second, we argued that postsecularity is commonly mobilised as a form of active resistance against neoliberal politics of subjectification. Mundane but subversive practices of care that challenge the logics of deservingness, eligibility, and the conscious cruelty of welfare reform, combined with the dissident public voice of some religious theo-ethics, has delineated a thirdspace where religious and secular publics can coalesce to form a 'counter-resonance machine' (Connolly 2008). Third, postsecularity cuts through the affective condition of ennui and fear by generating collective energies of hopefulness; not of mediocre optimism but a deeper capacity to believe in what currently seems impossible. Finally, we offered postsecularity as a mode of post-disenchantment. We suggest as diverse groups enter into dialogue with one another alongside their initial response to need, new subjectivities and more fulfilling notions of enchantment are emerging that have the potential to liberate people from neoliberalism in a spiritual sense, offering routes for conscientisation, and generating a potential basis for political and material change. Through dialogue with the other, desire and collectivity begin to be negotiated, drawing different pre/political motivations into a process of mutual transformation, highlighting the unsatisfactory 'enchantments' of neoliberalism – and certain forms of religiosity – and forging ahead to new connections between personal belief and collective action.

Attentiveness to the connections between belief and action inherently questions the interface between faith and reason, religion and secularism, and mystery and structure, thus affirming subject formation in neoliberal landscapes of need as an intrinsically postsecular matter. Even secular political action rests on animating enchantment – affectively underpinned commitment to a cause – but postsecularity underscores that for this to be progressive, this enchantment must be interrogated through a non-binary lens that engages desire in negotiation with collective action rather than simply being drawn back into a neoliberal grid of gratification. We illustrated how this postsecular subject formation is functioning, and the politics it can generate, through three examples. First, we illustrated the politics of receptive generosity at work in radical faith-based initiatives providing places of respite and care for highly stigmatised populations. The example of the Pauluskerk in Rotterdam revealed an opening out of faith-oriented space to make space for the other, a process that generated more compassionate approaches to substance misuse and re-scripted the dominant ideological meanings of public space during a time of zero-tolerance. Second, we highlighted how volunteer-activists in the Calais 'Jungle', are finding new ways of connecting belief and action through spiritual modalities, blurring boundaries not just between personal desire and collectivity but the appropriateness of universality versus localism in political action. Through a geography of encounter, consci-entising processes were shifting people's subjectivities, altering not only polit-ical imaginations beyond the space of the Jungle – shifting from a response to need, to advocacy and demands for structural change – but also the politics of the space itself, emphasising that it is not just a space of need but one of reim-agining society. Finally, we highlighted the new enchantments emerging in Christchurch after the 2010 and 2011 earthquakes. As both secular and religious institutions responded to need in the aftermath of the earthquakes, they joined together in order to imagine a new Christchurch, that posed the possibility of moving past the neoliberal status quo that preceded the earthquakes. Through a common ethic of co-producing a sense of place, different prepolitical disposi-tions could be renegotiated together, blurring the supposed religion/secularism boundary and generating a collective sense of the *soul* of the city being marked by in-commonness and grassroots experimentation. This was done through various artistic, environmental, and social justice projects that emerged in the gaps in the city generated by the earthquake and bubbled up through these cracks to overrun, re-enchant, and re-configure the affective geography of Christchurch. Building on this analysis, we drew on recent events in Charlottes-ville and Grenfell to demonstrate the importance of event-thinking in revealing new ethical, political, and theological subjectivities and spontaneous affinities and solidarities across diverse religious and secular groups that can rupture con-ditions of possibility.

In Chapter 4, we highlighted and analysed the *spatial assemblages* that are enabling postsecularity to bubble-up. Here we emphasised that in order to assess the ability for postsecularity to emerge in a particular place and to understand the dynamics of extant postsecularity, a historico-spatial analysis is necessary. Facilitating postsecularity requires proximity to difference, opportunities to encounter that difference, and resources – such as leaders, time, forbearance, education, ethical sensitivity, creativity, capital, a receptive sense of place and identity – to create a fecund environment for renegotiating the boundary between secularism and religion. A reason to come together is also necessary. This might be a search for community, curiosity about the other, or simply as a response to need. All of these things are crucial factors in generating the ethical proximity and emergence of in-common purposefulness and meaningfulness that character-ise postsecularity. These factors are thoroughly mediated by space. Attending to the spatiality of postsecularity emphasises its overlap with a feminist ethics of care which underscores the innumerable interrelations between people and all other materiality in the formation of subjectivities. This not only highlights the importance of spatial inequalities in analysing postsecular potentiality but under-scores the importance of the generosity and a non-servile humility present in the speculative formulation of ontology that is core to postsecularity. The sensitivity that postsecularity encourages towards the immensity of materiality enables people to demonstrate generosity through *performativity* and its inherent rela-tional instability (self-to-other *and* self-to-self); experimenting – in an ethically sensitive way – in encounters with difference by being creative with collective practice as well as spatial and material curation. We recognised that awareness of space, materiality, and performativity is crucial to understanding (and encouraging) the dynamics of existing postsecularity.

So, how can spatial analysis be used to increase ethical proximity and generate affective and political in-commonness? In this chapter we highlighted various settings in which responding to need alongside others with different motivations for action intentionally and generously taps into the transformative potential of performativity. Unique qualities of postsecularity emerge based on this performative engagement with difference and space. As various actors with differing religious and secular framings for action engage with one another in order to respond to injustice, the boundary between religion and secularism is renegotiated as religious people cede the position of ontological or metaphysical primacy and those invested in secularism recognise benefits that religious actors can make to the public sphere that they have not recognised before. We argued that this postsecular generosity has generated new subjectivities and ways of being in-common that have enabled groups with motivational differences to tackle issues such as climate change; coming together in liminal spaces where people have untethered themselves from the status quo of their usual polity and entered into a subjective transition of values and identity. As religious people have recognised the shortcomings of their institutions – both organisationally and discursively – in tackling climate change, so those from the scientific community have begun to appreciate the usefulness of spirituality in motivating action on climate change. From both secular-scientific and faith-motivated perspectives, practices of rapprochement can in turn bring about tensions emanating from institutional fundamentalisms; 'green religion' and 'spiritual science' sometimes struggle to be taken seriously within their original institutions. However, their appreciation of one another and their collective work through movements such as Transition Towns begins to renegotiate traditional boundaries between faith and reason, and religion and secularism, generating communitas; a community of marginalised people bound together by their common journey of transition away from their given identities and values. We argued that liminality and communitas are generated by spatial context and curation. We used the examples of Hope House (a homeless shelter, and drug and alcohol centre run by the Salvation Army) and Levington foodbank. Analysing these spaces through our three modalities of postsecularity (rapprochement, re-enchantment, receptive generosity), we illustrated how they exhibited postsecularity as a practice of liminality. These organisations were informed by their specific spatial context but then generated communitas, hence producing values in-common with those that they worked alongside as part of rejecting or subverting their initial *raison d'être* through a process of mutual transformation and liminal transition. In the case of Hope House, the organisation subverted neoliberal governance by forming a common ethical practice between its diverse staff that rejected metrics of deservingness and disciplinarian ways of dealing with its clients in order to provide a service modelled on unconditional support. Micro-practices such as eating together, alongside the performativity of staff and residents, produced an affective sociality that expressed receptive generosity and opened out liminal spaces of encounter, albeit temporary. At Levington food-bank, volunteers transitioned from motivations informed by apolitical forms of Christian charity towards a politicised position of advocating against punitive government measures that exacerbated poverty and food insecurity. In showing generosity to one another's worldviews, the staff, volunteers, and clients in these organisations enacted a liminal politics enlivened by their common performativity of the spaces of action. This generated a resistant politics and a new in-commonness against injustice, renegotiating prepolitical predilections and blending religious and secular ethics to generate new political subjectivities.

In Chapter 5, we analysed how postsecularity can be identified within and facilitate political practices. Here we suggested that postsecularity broadens the ontological scope of progressive politics beyond traditional Left/Right divisions and opens up political practice to greater – but sometimes risky – hopefulness, strategic performativity, ethical sensitivity, and continuity between belief and action. We argued that it is difficult to enact a politics that places hope in the Other, but that an ethos of postsecularity embraces the inherent fallibility of embodying this hope. Embracing this hope renders the individual or collective vulnerable to exploitation, but can also unleash unforeseen political empower-ment, generate new allegiances, and exhibit fruitful tactical agility. Through our three modalities of postsecularity, we asked what practices can help to generate generosity towards the Other and then what additional practices can convert that hopeful generosity into political action. By first addressing the modality of re-enchantment, we emphasised that an affect of openness to the Other is crucial to postsecularity. Hopeful, animating enchantment and ethical proximity are missing from many political movements, but can be found emerging in arenas such as hospitality and community organising, and can be facilitated by the prac-tices we highlighted in the rest of the chapter. However, we argued that the prac-tices that facilitate openness to the Other often emerge in more mundane and creative spaces rather than overtly political ones. We illustrated how participa-tory art in marginalised communities can lower interrelational defences and generate bonds of enjoyment, trust, and shared pain which increase the ethical proximity of participants. This closeness can lead to recognition of one another's dreams, disappointments, and feelings in-common, which – particularly in mar-ginalised groups – can emerge as new affective topologies that cohere around political issues. For instance, in Finley and Finley's (1999) work, the New Orleans poetry scene is highlighted as a medium through which homeless people could share their art with others, generating a broader movement beyond the homeless community that raised awareness and directed critique at the city's regressive political and social culture regarding homelessness. We highlighted that postsecularity in this instance has serendipitous political consequences, the art spaces being designed for participation and inclusion rather than for deliber-ate political ends. However, we also noted that some religious artists are using participatory art in order to generate postsecularity with progressive political outcomes. Through Moody (2012) and Ganiel's (2006) work, we examined the example of the Ikon community in Northern Ireland and discussed how they ground speculative ontologies so as to ask existential questions about how meaning and religious authority are legitimised. Ikon's participatory art practices encourage people to conceive of Christian faith as a hope that God can be encountered through ethical negotiation rather than blindly obeying rules handed down from institutional biblical interpretations. By applying this lens to particular issues such as women's liberation or the Troubles in Northern Ireland, Ikon unsettle the boundaries between religion and secularism, stressing the importance of collective engagement in public life and the possibility of religious experience emerging in the midst of messy spaces filled with mixed motivations and variegated metaphysical predilections. The speculative ontology they seek to embody generates a community bound by disavowal of the religious and social status quo and waits for new, enchanting – perhaps divine – meanings to emerge and inspire collective action.

After highlighting the affective conditions needed for postsecularity – common grief/rage/joy followed by emergent topologies of enchantment that embrace the Other – we examined the second modality of receptive generosity which we defined as a process of iterative reflection upon and ethical evaluation of relational friction in collective praxis. We illustrated how religious communities foster techniques of self that foreground reflection on connections between self-to-self and self-to-other relationships, facilitating their members' reflexive praxis and often shifting it towards political activism. This is done by intentionally making space for community members to exercise reflexivity (through discussion or meditation – or through rituals – like sung worship, prayer, and ceremony) which have the performative potential to unsettle religious subjectivities, encouraging reflection upon the source of discomfort. The religious communities we analysed performed an ethos of postsecularity by encouraging members to question religious certainty and facilitated encounters with difference, supporting their members as they incorporated these experiences into their praxis. These practices – questioning certainty and encountering difference – helped religious people to think deeply about what they believed and how that related to ethical action. This created an environment that empowered members' political activism at both individual and collective levels, creating affinity groups within their faith community or joining/starting political causes in the heterogeneous public sphere.

Finally, we examined how postsecularity was exhibited through rapprochement between Christians and other activists in the Occupy movement. Here we emphasised that rapprochement demonstrates a more hard-nosed approach to renegotiating supposed religious/secular divisions. Often, rapprochement emerges from a pragmatic desire to do something about something, forgoing difficult discussion about metaphysical and motivational differences in order to generate a tightly focussed set of narratives about what the interested parties have in common and want to achieve. However, we argue that in the Occupy movement the strategic intentionality of rapprochement transformed the movement into something more radically postsecular. Because rapprochement creates a buffer-zone around the goals of a collective, differences can be explored without the fear they will jeopardise the movement's *raison d'être*. This enables new subjectivities, tactics, and political imaginations to emerge. In the case of Occupy, the movement developed its focus, building on its anger at the neoliberal status quo to generate crossover narratives inspired by religious and secular resources that questioned economics as a meaningful metric for planetary well-being and championed horizontal democracy. This chapter, then, suggested a series of different practical ways in which postsecularity can be performed, and argued that its successful facilitation requires activists to be spatially and ethically attuned. Furthermore, it foregrounded examples of public seizure of the psycho-spiritual terrain occupied by neoliberalism; demonstrating the contemporary political contributions offered by postsecularity, both in terms of new ontological approaches that incorporate both desire and ethics, and with regard to the potential of spiritual activism as a significant element of participatory action research.

In Chapter 6, we began by acknowledging that much of our analysis of postsecularity had been focussed on religious-secular (re)negotiations in the Western European context. We wanted to use this chapter both to highlight the conceptual utility of postsecularity at a more global scale, and to respond to concerns that the concept has limited relevance beyond the Christian-secular nexus of engagement in the West. By reviewing key debates and literatures, we sought to provide a layered analysis of the spatial formation of postsecularity in non-Western and predominantly non-Christian settings. We argued that even though the global mission of secularisation has failed and that many governments still exhibit theocratic tendencies, the performative aspects of secularism linger in the political structures of most post-colonial states as well as in societal imaginations. In this sense, the ongoing postsecular negotiation between secular structures and religious sensibilities is arguably a global issue. Equally, through a series of case-studies we demonstrated how the co-constitutive dynamics between religion, belief, and secularity are becoming increasingly blurred and hybridised, presenting distinct opportunities, and barriers, for the emergence of postsecularity. First, we discussed gender politics and highlighted the emergence of postsecular feminism as a critique of both the Eurocentric Christian 'modern' subject, and the gendered and racialised exclusions associated with hegemonic imposition of Western secularism (Braidotti 2008; Mahmood 2005; but also see Vasilaki 2016). The renewed lines of generosity and openness towards religious subjectivities within feminist movements, and the recognising of religious sources of agency, is highly pertinent in the contemporary resurgence of political conservativism and explicitly anti-feminist, Islamophobic ideologies propagated by politicians. We argued through Göle (2010) and Gökarıksel and Secor's (2014) work that Turkish women exhibit elements of postsecularity by renegotiating different politically charged threads, and through the role of veiling constitute themselves as 'ethical, pious and desiring subjects' (p. 95), thereby seeking to trouble the boundaries between religious and secular claims to authority on their lives. Second, we gave examples of the growing trend towards interfaith collaborations in international development and humanitarianism. While questions about the place of international NGOs in local state governance, sovereignty and territory remain, we pointed to examples of receptive generosity and rapprochement across lines of religious and secular difference, suggesting a weakening of religious and secular fundamentalisms. Third, we examined the ways in which different types of state-religion relationships, combined with changing global geographies of secularity, religion, and belief, generate highly uneven possibilities for capacities for postsecularity. We highlighted the ambiguous role of the Russian Orthodox Church in the new Russian nationalism, the postsecular symbolism and subversive protests of Pussy Riot, as well as the growth of Islam and Pentecostalism which have taken up key welfare provision responsibilities and partnerships with the local state. We discussed the contentious politics of Indian secularism, and the localised and material geographies that secure everyday peace between Hindu and Muslim populations (see Williams P 2015). We then addressed the changing geographies of Chinese and Hong Kong secularisms and how each is undergoing intense negotiation with regard to the place of religion in society, before tracing signs of postsecularity in the Hong Kong Umbrella Movement. Last, we drew attention to the sensory and material geographies of religion in urban spaces, and the ways in which these shape everyday practices of enchantment, conviviality, tolerance, and (in)difference – all of which point to a locally contingent and contested terrain on which ethical capacities of receptivity and rapprochement emerge.

These new multi-layered and multi-spatial readings from more globalised contexts of postsecularity, highlighting as they do the constantly shifting and evolving relationship between religious, spiritual, and secular imaginaries, inevitably challenge some of the intersectionalities of postcolonial theory where until recently, religious identities and subjectivities were largely bracketed out. Rather, new understandings of agency and autonomy emerge in which essentialised readings of both the religious and the secular are challenged, and tired and outdated imaginaries of the nation state as the only vehicle of postcolonial aspiration are critically assessed. More needs to be done to develop the fruitful challenge presented by Ratti, and his call for a new postsecular postcolonial imaginary, in which fresh conceptions of secularism and religion can emerge to potentially generate new forms of ethics that will reject the national violences of the past. The challenge is to now progress from his distillation of this new imagination from different forms of postcolonial literature, to one that is embodied in material imaginaries and practices of postsecularity that are generated from those new forms of postcolonial (re)enchantment and rapprochement between different ontologies and actants that we have begun to identify in this chapter.

7.3 Hopeful and non-binary ethico-politics

Having examined the ways in which this book has illustrated that postsecularity presents a progressive ethos for renegotiating the interface between religion and secularity, we now turn our attention to positive interventions that we think that postsecularity can offer to political activism and the geographic academy. These possibilities refer to four areas: the relationship between affect and politics; the contribution of spirituality to activist praxis; new maps for political analysis and action; and new ontological, epistemological, methodological, and empirical approaches to geographical research.

The relationship between affect and politics

Here we want to consider the possibility that postsecularity is interconnected with new forms of *alternative affective politics*. Sitrin's (2006 and 2012) notion of 'politica afectiva' emphasises how alternative political responses can be fostered through the creation of a base that is loving, supportive, and built on trust. As Woodward (2011) puts it, we need to be aware of the affective dimensions that figure in the production of a dynamic environment of collaborative social relations and collective action. Recognising emerging politics of social relations and love will require attention to how subjectivities and spaces shape, and are shaped by, these affective dimensions. First, we have noted the capacity of participants to affect and to be affected by others, for example, in the phenomenology of need that arises from working alongside vulnerable others in an atmosphere of collective trust. Such experience affects participants and offers an affective politics into their participation. Equally, participants can themselves contribute to a more collective form of social responsibility by the ways in which they perform co-operation, mutual support, and in-common connection; a variation of what Hardt and Negri (2001) have termed affective labour. However sub-optimal the performance of these affective politics turns out to be in practice, the presence of alternative feelings, vibes, and atmospheres relating to non-hierarchical and collaborative ways of being, opens out possible ethical capacities for further alternative affection and recuperative instincts in amongst the seemingly mundane activities of care and responsibility found in the meantime.

Postsecularity above all recognises the importance of performativity in the political realm. Geographies of performativity often highlight the experimental nature of art practices which convene new self-to-self and self-to-other relationships. However, these relational experiments – which can be perceived to be ephemeral – have the potential to initiate lasting transformation. Furthermore, the relational experiments of performance are not contained within the artistic realm but are present in almost all sociality, including politics. In Chapter 5, we highlighted how the postsecular mode of receptive generosity was embodied by activists pursuing novel political enchantments in-common and collective *raison d'êtres* in Sutherland's accounts of ecumenical social justice conferences and Exeter Church. By deliberately experimenting with uncertainty-inducing political relationships and applying a reflexive praxis, performativity emerged as a vector through which progressive political subjectivities and movements could emerge (although – as is the nature of experiments – also being susceptible to disappointment). We argue that postsecular performativity, in whichever mode, exhibits an experimentalism that engages the Other and embraces the inherent potential that this interaction has to transform power relations.

In the current political climate – particularly in the West – of rising neofascist populism and neoliberal governmentality, employing a postsecular performativity of experimentalism may seem too risky, but we would argue that it contains the potential to wrest back control of the affective terrain of politics that analysts like Klein (2017) and White (2016) argue has been so effectively colonised by neoliberal ideologues. Klein in particular argues that the efficacy of contemporary figures such as Trump, Farage, and Le Pen lies in their ability to stoke up an atmosphere of fear and resentment. Rather than directing critique at the elites that caused the financial crises of 2008 for economic instability and falling living standards, these figures have redirected public anger towards marginalised groups, and painted policy-makers that show any vague sensibility for helping minorities as smug hypocrites who massage their consciences by posturing as the friends of outsiders whilst harbouring disdain for 'normal people'. These neofascist figureheads not only marshall the affective world of those who support them towards ressentiment but generate a constant flurry of mediated spectacles – which alternate between inflammation of culture wars and regressive policy statements – that chew up the affective energy of progressives through persistent outrage. Klein suggests that even though there are necessary policy interventions to be made to stem the tide of – particularly Trump's – political atrocities, a more necessary political goal is the preparation of a 'people's shock' (2017, p. 234). She suggests that what is needed is for a significant number of progressives to retreat from engagement in the seething back-and-forth of trying to out-mediate right-wing demagogues and instead to engage in a praxis of spiritually enriching and value-attuning direct action alongside difference. Her key example is the DAPL pipeline protests at Standing Rock which brought together a coalition of indigenous peoples, army veterans, environmental activists, faith groups, and a whole range of others. She argues that by binding together to serve one goal – protecting the water – a space of generosity opened up where ethical engagement, learning how to look after one another as a community in an outdoor camp, and witnessing one another's cultural and spiritual rituals and practices took place. This led to various experimental, performative encounters and the emergence – or communal outpouring – of collective values and desire. Performative practices that facilitated the emergence of new affective topologies included forgiveness ceremonies – where veterans were pardoned by indigenous activists for the American military's role in breaking land treaties – and communal participation in indigenous sage smudging rituals. These performances not only facilitated mutual transformations across the group, reconfiguring self-to-self and self-to-other relations, binding difference more tightly together under the initial cause, but led to a new affective sense of what the group had in common beyond the DAPL camp. Klein points to the Leap Manifesto, which emerged out of the DAPL action and moved conversation beyond a simple 'no' to the violations that the Trump administration is wreaking upon its own citizens (and the rest of the world), to being daring enough to be enchanted by a radically alternative vision. Klein states that, more than a list of policy directives, the manifesto creators tried to operate from a foundation of spirituality, values, and enchanting desire, that embodied love for others and the planet, producing a surprisingly detailed set of policy directives focussed on land rights for indigenous communities, clean energy, public infrastructure investment, sustainable agriculture, immigrant rights, and an end to austerity.

We argue that this new sense of enchantment and new vision of the world emerged from the practice of postsecularity, a performative embodiment of generosity leading to subjective transformation and the emergence of progressive affective topologies and renewed enchantment that stands in stark contrast to an affective landscape of ressentiment, anxiety, and ennui. But how can this sense of enchantment supplant the affective terrain generated by the current neoliberal conjuncture on a grander scale? White (2016) argues that through persistence of the spiritual conviction generated by an ethically attuned and relentlessly praxis-based sensitivity for emergent enchantment, revolutionary moments emerge in history where the self-contained spiritual power or wholeness of progressive enchantment sweeps across society, transforming it wholesale. But society has to be ready for this, it has to be prepared; this is why postsecular praxis is important. Citizens engaging in a praxis that is generous to difference and unleashes desire in an ethically attuned way through collective praxis is capable of generating broad-based resonances and has the power to confound the previous order politically by highlighting its absurdity and obsolescence. Therefore, we posit postsecularity as a kind of prefigurative revolution-in-waiting, generating not only new small-scale affective topologies that are capable of winning progressive victories on single or local issues – which slowly erode the legitimacy of and close the net around worldviews based on ressentiment – but preparing the ground for a broader, revolutionary enchantment of networked resonances based on generosity towards the Other. Again, rather than a simple 'no', we see postsecularity producing more specific visions of how the world should be; from the participative democracy of Occupy, to the championing of religious liberty in new hospitality movements, to the generation of inclusive creative cultures in homeless arts practice. We are hopeful that as the spiritual and practical bankruptcy of neoliberalism becomes increasingly apparent – along with the concomitant horror that accompanies a realisation of its fundamental inability to support a sociality that can meet material and emotional needs – a growing preponderance of postsecular activity can prepare the way for a connectivity of spiritual resonances to sweep through society, connecting desire with ethics and translating an enchanting vision into practice. We argue that this spiritual mode of affective politics can reinvigorate the Left and generate more effective opposition to the Right, which we turn our attention to next (see Gilbert 2017).

The contribution of spirituality to activist practice

Our second suggestion regarding the contribution of postsecularity to politics in the future is that in renegotiating the boundary between religion and secularism, it enables activist communities to draw more effectively on both religious and non-religious spirituality. As postsecularity reworks the boundaries between religion and secularism, an appreciation of the separate roles of the two emerges; religion as a posture that hopes for the in-breaking of seemingly impossible ways of being and seeing, and secularism as a political system that ensures that that process of existential exploration is open to as many people as possible by ensuring no singular way of being and seeing is allowed to dictate to the rest of a polis how they should be and see. We argue that postsecularity works effectively in tandem with a spirituality or 'spiritual activism' which sits between religion and secularism as an a/theistic way of being that embraces the transcendence of existence. Spiritual activism is embodied in the recognition that whether a subject has religious or irreligious predilections – between their self, their desires, and their social modes; there are significant gaps in knowledge and tricky ethico-political quandaries regarding the interrelations between desire, self, and sociality to navigate which have no clearcut solution but rather demand strenuous negotiation and reflexivity (see McIntosh and Carmichael 2016). This often requires acknowledging the most traumatic elements of existence that many political commentators might refuse to fully account for, especially the futility of social movements without a soul; that is, activists that are not drawing their inner empowerment from enlivening psychodynamics. Postsecularity actively explores the trauma, desire, and gaps in being and seeing that spiritual activism highlights and generously experiments with Others in these grey areas in the hope that new affective topologies and identities in-common emerge. We argue that this postsecular spiritual activism offers two benefits to political activists.

First, spiritual activism can reinvigorate the Left itself. As right-wing populism rises at the moment in the West, the Left is troubled both by a sense of listlessness as precarious conditions sap the capacity to fight back, and by divisive ideological defensiveness as anxiety about societal and environmental collapse mounts. Given the evidence presented in this book, we consider that spiritual activism can help the Left by transcending anxiety and reframing radicalism, creating generous parameters for a broad interconnection of revolutionary movements, and promoting egalitarian modes of organising. As we highlighted in the case of the Leap Manifesto, postsecular movements aim to generate and explore new routes-towards and qualities-of enchantment. Spiritual activism adds to this by underscoring, in the vein of speculative ontology, the opacity of a Master Signifier that might buttress any sort of status quo. It therefore aids a recognition that any sense of enchantment can only be constructed when connected to the existentially tangible entities of desire, self, and other. In particular, spiritual activism emphasises that movements that do not enliven the psychodynamics of desire within their participants lead to burnout, resentment, disappointment, and cynicism that paralyses any attempt at activism. By highlighting the importance of desire, the concept of spiritual activism gives theoretical permission for activists not to go searching for the most supposedly politically astute form of activism or organisation to belong to, but rather to unleash the creative capacities of their desires in conversation and ethical proximity with others. This, of course, means that desire-inspired activists will be dispersed into a variety of different issues and scales. However, we argue that this generates a postsecular way of operating, generously and experimentally engaging with others in order to figure out the best way for the activist to release a creative and joyful politics, and generating broader parameters for enlisting political engagement, giving rise to new forms of enchantment and political consciousness. Spiritual activism offers a wider political platform to stand on, where it is more likely for a plethora of movements to connect-up and change the status quo by recognising, as they experiment through postsecular performativity with others, the deeper desirous resonances that they can generate together. Additionally, in this state of desirous creativity and generous engagement, activists are more likely to take increasingly radical steps to alter the status quo, being possessed by an enchanted politics of love, that engages in self-sacrificing actions – seen in the Zapatista movement, the Suffragettes, the Civil Rights movement, and ongoing direct action against the preparation for war – not out of duty or institutional subjugation, but out of a new consciousness that is bewitched by the possibility of a new world bursting forth. Furthermore, as spiritual activism honours the truths that are difficult to accept – the horrendous realities of the world that many people try to screen out – it resonates with the postsecular desire to include the most marginalised as central parts of the movement, acknowledging the great spiritual wisdom that they have to impart. These groups have often suffered in such a way as to need to find enchantment in the most oppressive of circumstances. It is necessary for their voices to be heard in postsecular and spiritual activism, not only to teach people how to find empowerment when fighting an imposing hegemony but also to highlight when activists have settled for a hope that is in part based on the status quo, settling for cynical exclusions that emerge from self-pity rather than the deep animating enchantment that emerges when all hope is seemingly lost. This powerful hope beyond hope generates a radical vision of a new world that alters oppressor and oppressed, not simply targeting modest reforms that satiate moral offense.

Second, spiritual activism sharpens political sensibilities of postsecularity by providing a stringent riposte to 'dark' and 'heavy' forms of the postsecular. Postsecularity renegotiates the boundaries between religion and secularism, but some of the most prominent renegotiations between religious and secular forces are very toxic assemblages indeed. For instance, the Trump administration relies heavily on the support of white evangelicals and in the UK, fascist groups like Britain First have plied the narrative that Britain is a 'Christian nation' in order to stoke anger towards immigrants who purportedly threaten this religious national identity. Although these movements show some 'generosity' in accommodating both religious and secular predilections so they can act together, we argue that these crossovers, although illustrating shades of postsecularity (in their renegotiation of the religion/secularism boundary) do not embody postsecularity in a way that unleashes its most constructive powers. These overlaps do not generate the openness to subjective transformation that postsecularity works towards; but rather, embody a cynical tolerance that quickens the construction of hard divisions typical of ressentiment. These overlaps shut down the emergence of alternative regimes of desire and foreclose political debate that addresses the most painful issues in order to set up 'functional' solutions, something postsecularity and spiritual activism are diametrically opposed to. Although postsecularity elicits a radical generosity that engages in experimental relationships – even with problematic others – in the hope that new ways of being will emerge, spiritual activism can help those practicing postsecularity to discern when these relationships are straying towards 'heavy' or 'dark' postsecularism. Postsecularity informed by spiritual activism undermines the resonances between conservative religious and political elites by more deeply interrogating the interface between religion and secularism. As it defines the roles of religion and secularity, postsecularity highlights the grey areas in knowledge that exist between them and the impossibility of metaphysical primacy. Spiritual activism triangulates these grey areas more specifically within desire and ethical engagement. The way in which dark postsecularity crushes, defers, and deforms any desire that is alternative to its regime – a self-evidently unethical practice (see Levinas 1978) – is anathema to postsecular spiritual activism. Spiritual activism argues that the things that bind people together most effectively are shared ethical proximity, nearness to their own and Others' pain, and celebration of collective creative emergences that are ethically tested. Postsecularity, aided by spiritual activism, has a sensitive ear for the suffering of crushed desires and for the hope which wells up from the hopeless. It renegotiates the secular/religious boundary in order to better listen for these voices, to acknowledge the legitimate ways of being-in and seeing the world that the status quo censors. The counsel that it provides to postsecular activists is that anything which dulls this sensitivity for the suffering other must be treated with caution, helping activists discern how regressive forces might try to take advantage of the generous tenderness towards the pain of even problematic others that is available in postsecularity. Spiritual activism can help steer postsecularity's polyvocality away from the 'raw opening of possibility' towards social justice.

New maps for political analysis and action

Third, postsecularity offers an alternative cartography for activist analysis and action. We argue that the modalities that we have illustrated throughout this volume – rapprochement, receptive generosity, and re-enchantment – can be moved between and tactically applied. As we discussed in Chapter 4, the potential for postsecularity is contextualised spatially, affectively, and performatively. The ways in which different modalities and performativities of postsecularity can be shifted between can help activists to read and influence a spatial/affective/performative context and its postsecular potential, facilitating the liminal spaces and their emergent communitas that trigger the becoming of new affective topologies and identities that characterise postsecularity. Explaining how these different modalities can be moved between can help activists to facilitate the emergence of the other-embracing topologies of affect and collective con-sciousnesses that postsecularity aims for.

As we have illustrated in our case studies, postsecularity emerges from liminal spaces, where extant meanings are suspended, and new collective mean-ings are generated, forming a new community or 'communitas'. This loss of meaning may be an existential or contextual condition but may also be brought on by rapprochement – dialogue with the Other in order to find a pragmatic way of acting together – or through ethical proximity with the Other through altern-ative performative practices such as running a foodbank, attending a protest, convening an art workshop, or participating in environmental activism. This meaninglessness is transcended by bringing the ethical concerns of a group – their desires and disappointments – into conversation and finding commonality; common values, dreams, and affects that can be operationalised so as to manifest a collective desire. This can result in an affective re-enchantment that generates new ways of performing community but can also spill over into the political realm, confronting injustice with collective action. Once this new way of being together is in motion, it can be experimented with, activists can test what else is collectively possible through the exercise of receptive generosity, reflecting on areas of relational tension, and thinking up practices that might help them tran-scend relational tensions. This might involve moving back through different postsecular modes, whether that be rapprochement – agreeing for a while to dis-agree on something but bearing with one another to see that might change – or performances that generate new kinds of ethical proximity, whether that is spaces of ritual (as in religious communities), conviviality, or creativity, giving rise to new forms of enchantment. All these transitions are bi-directional and can be shifted between and/or stayed within depending on the need of a collective, its facilitators, capacities, and the goals they want to attain.

New directions for geographical research

To take this agenda forward, therefore, we present four new avenues for geo-graphic research. First, we call on critical geographers of neoliberalism and political-economy to interrogate the highly contested relation between ethics, affect, and spirituality. Reconceptualising neoliberalism upon the affective regis-ters of the personal, the aesthetic, and relating to desire, we have argued, re-vitalises the study of late-capitalist subjectivities and the ways it is shaping religious and nonreligious belief and practice, but also foregrounds the potential spaces of resistance. Understanding neoliberalism as an affective politics that shapes ethical relations is not new *per se*; yet the contestation of neoliberalism through spiritual activism is rarely understood. We have developed a set of con-ceptual tools of precorporation, ressentiment, and enchantment to diagnose the subtle ways in which neoliberal logics shape the formation of ethicality. Spiritual activism, we argue, is a key practice that helps name and excise the shadows of neoliberal and capitalist desires that dwell within us and perpetuate an estrange-ment and vindictiveness towards others. Geographers have much to contribute in mapping the spatial formation of the variegated ethical and political subjectivities arising as traditional categories of the secular and the religious become enfolded and deterritorialised in different directions. We suggest Rose's notion of post-disenchantment is useful here to frame the ways in which late capitalist subjectivities are entangled in contested processes of enchantment and disenchantment. While the concept was developed to question the helpfulness of seeking more 'enchantment' given contemporary capitalism is intent on tightening the bond between consumerism, desire, and alienation (Rose 2017), it has been deployed here to interrogate the divergent politics of enchantment at work in secular and religious belief and practice, highlighting their hopeful as well as right-leaning manifestations. We suggest enchantment remains politically charged despite the attempted precorporation by capital, and that postsecularity and a/theistic spirituality might enable activists to discern a progressive route through the murky waters of desire. This raises a series of pertinent areas of research embracing questions about the practices used by activists to nurture and harness desire, and the performances by which they enliven such desire amidst neoliberal regimes that seek to precorporate and deform their energies. Such questions have particular relevance in connecting recent discussions on spiritual ontologies amongst geographers of religion with mainstream political geography. They are also highly relevant for discussions which link the political with a deep secular enchantment, such as in the discussion of *Cosmopolitics* (Blok and Farias 2016) that posits speculative realism as the basis for a new 'ontological politics' that can galvanise new ethical practices for materialising the commons. Additionally, we suggest one area of critical importance concerns the capacities that are unleashed when activists reflect upon their overlaid maps of political and religious subjectivation (see Häkli and Kallio 2014; Sutherland 2017). This builds on Holloway's (2003) work in the geography of religion, which highlights the importance of spatial context to religious experience, but also the reflexive capacity of the subject to understand, react to, and alter their spirituality. However, postsecularity begins to generate new lenses on political movements that render visible the ways in which some elements of religion and spirituality generate changes in consciousness, resulting in alternative psychogeographies. For instance, what spiritual techniques – altering the power relations between self-to-self and self-to-other – do political movements such as the Zapatistas or the DAPL movement use that so radically cut them out of the hegemonic consciousness of the time, creating a new collective consciousness that is willing to take radical risks in the fight to imagine another world? What are the spatialities that make this possible? These questions highlight the inherent usefulness of postsecularity in questioning the boundaries between religion and secularism. What political ontologies does this querying allow to come to the fore, and what kind of activism do they empower? As new, more speculative ontologies potentially arise out of postsecularity, how can the haziness this creates between ethics and politics, faith and reason, mystery and structure, be progressively responded to? Exploring the contested relations between enchantment, desire, and hope therefore are highly pertinent questions for geographic consideration.

Second, one of the important contributions of this book has been to explore the nascent and emergent permutations across the contemporary religious and secular landscape of belief, ethics, and subjectivity. We drew attention to the contested arena of disenchantment and enchantment shaping affective life, ethics, and subjectivity. It is time for geographers to pry open the 'black box' of unbelief and examine its varied existential cultures (Lee 2015 and 2017), recognising that categories of the sacred are not the preserve of the religious but are diffused throughout everyday life. This book has contributed to this agenda by highlighting new forms of 'sacralisation' and re-enchantment that are crystallised in anti-poverty and environmental activism, opening out possibilities for rapprochement with socially engaged religious activists. There is still much to learn about secular voices and desire in spaces of rapprochement. This is essential if we are to understand the variegated expressions of postsecularity. This would entail analysis of *predispositions* – that is, the political and ethical norms and visions for change that secular citizens already bring into the public sphere; *motivations* – on what basis do secular individuals and institutions enter into postsecular partnership and what do they perceive as the advantages or drawbacks to such an initiative; *tactics and negotiations* – what approaches are adopted to engagement and joint-working with religious or faith-based individuals and organisations, especially when working with socially marginalised people and other key ethical and economic issues associated with neo-liberalism; *practices* – what roles do secular citizens play in leadership, organisation, and the fashioning of local democratic and participatory processes and where are the points of resonance and dissonance between secular and religious worldviews in practice and how are these negotiated; *transformations* – what changes in cognitive and ethical worldviews of individuals occurred as a result of these partnerships; *impacts* – in what ways do postsecular partnerships contribute to shaping the debates and practices of civic and political engagement, and what extra value is perceived to have been added by these partnerships, for example in new forms of democratic solidarity, tolerance, hospitality, and what hindrances are also evident?

Third, formations of postsecularity are grounded in discourse, practice, and materialities that can circumscribe and curate progressive potentialities. Throughout our case-studies we have sought to develop an analytic of postsecularity that is empirically sensitive to the multiple and inadvertent ways that postsecularity is entangled in wider (geo)political relations and power dynamics. Taking forward this analysis is crucial if we are to avoid possible co-option of postsecularity within exclusionary practices. For instance, as postsecularity appears in the rapprochement of religious and secular organisations working in the field of anti-trafficking, resulting in prominent faith-based groups adopting a human right-based agenda, Lonergan *et al.*, (forthcoming) highlight the importance for geographers to critically examine the discourses and practices that underpin rapprochement. In some cases, rapprochement can be embedded within ethically problematic representations of sex trafficking that obscure the prevalence of slavery and human trafficking within numerous industries, and reinforce experiences of stigma and exclusion for trafficking victims and survivors (Zimmerman 2011; Campbell and Zimmerman 2014).

In addition to tracing the discourses and practices that organise spaces of postsecularity, it is important for future work to ground an analytic of postsecularity in the polyvocal experiences of rapprochement – resisting the temptation to 'read off' the experience and intention of the core members and recognise spaces of rapprochement can be perhaps more ambivalent spaces for particular groups and identities. Analysis of *actually existing* postsecularity therefore requires taking as central concern issues of marginality, identity, and intersectionality. This would entail a sensitivity not only to power-relations but also to marginal experiences within practices of receptivity and rapprochement, offering an analysis of postsecularity as experienced through the unique intersections of age, class, race, ethnicity, gender, sexuality, religion, (dis)ability, and subculture. Within this, the focus must be on the 'mutually constitutive forms of oppression' (Hooks 1982; Hopkins 2017, 1), including longstanding processes of racialisation and sexism, that might remain invisible to some yet produce asymmetrical burdens for others. Specifically, we argue the need to give central attention in spaces of postsecularity to the enchantments and spiritualities of marginalised people themselves. Receptivity needs to be dialogical to be meaningful and if it is to avoid reproducing the shallow enchantments discussed in Chapter 3 and 5. Without this, notions of in-commonness will remain politically limited.

Fourth, we have sought to draw attention to the different spatialities of postsecularity, acknowledging the spatially contingent and context-dependent character of receptive generosity, rapprochement, and hopeful enchantment. Research on postsecularity hitherto has predominantly been developed in European and North American contexts, and with regard to particular arenas of investigation such as urban marginality, welfare, and protest spaces. What might be learnt about the distinct capacities of postsecularity in different arenas (for example, education and healthcare), or spatial contexts (for example, the diverse and fast-changing religious and political geographies shaping African urbanism). Geographers are well placed to produce empirically sensitive accounts that trace the ways in which postsecularity works differently in relation to varying historical projects of secularism and changing forms of secularity and religiosity including post-atheist, polytheist, monist beliefs and subjectivities. The diverse and changing global geographies of religion, belief, and neoliberal political-economy present exciting opportunities for studies of postsecularity to catalyse wider religious and secular traditions to overcome a divisive politics by developing more hopeful in-common sensibilities. Examining the ethical capacities emerging through practices of receptive generosity, rapprochement, and re-enchantment in globalised settings, and exploring their divergent political possibilities, we suggest, is a fruitful agenda for geographers to pursue.

Bibliography

Ackerman B (1994) *The Future of Liberal Revolution.* Newhaven, CT, Yale University Press.

Adamson J (2017) *The City That Fell To Its Knees.* Independently published, Christchurch.

Agamben G (1998) [1995] *Homo Sacer: Sovereign Power and Bare Life.* Stanford, CA: Stanford University Press.

Agamben G (2005) *The Time That Remains.* Stanford, CA: Stanford University Press.

Ager A and Ager J (2011) Faith and the Discourse of Secular Humanitarianism. *Journal of Refugee Studies,* 24(3), 456–472.

Ahmad M and Rae J (2015) Women, Islam and Peacemaking in the Arab Spring. *Peace Review,* 3, 312–319.

Al Qurtuby S (2013) Peacebuilding in Indonesia: Christian–Muslim Alliances in Ambon Island. *Islam and Christian–Muslim Relations,* 24(3), 349–367.

Aldridge A (2000) *Religion in the Contemporary World: A Sociological Introduction.* Cambridge: Polity Press, 2nd Edn.

Alinsky SD (1971) *Rules for Radicals: A Practical Primer for Realistic Radicals.* New York, NY: Random House.

Allahyari R (2000) *Visions of Charity: Volunteer Workers and Moral Community.* Berkeley, CA: University of California Press.

Amin A (2006) The Good City. *Urban Studies,* 43, 1009–1023.

Anderson B (2012) Affect and Biopower: Towards a Politics of Life. *Transactions of the Institute of British Geographers,* 37(1), 28–43.

Anderson B (2014) *Encountering Affect: Capacities, Apparatuses, Conditions.* Abingdon, Oxon: Ashgate.

Anderson B (2016) Neoliberal Affects. *Progress in Human Geography,* 40(6), 734–753.

Anderson B and Harrison P (2010) *Taking-Place: Non-Representational Theories and Geography.* Farnham: Ashgate.

Anzaldúa G (2012) *Borderlands/La Frontera: The New Mestiza.* San Francisco, CA: Aunt Lute Books.

Arab American Tribe (2015) *The Economic Approach: Cooperatives and Coexistence as a means to Peace in Israel and Palestine.* Available from https://arabAmericantribe. wordpress.com/2015/09/24/the-economic-approach-cooperatives-and-coexistence-as-a-means-to-peace-in-israel-and-palestine/ (last accessed 27/07/18).

Asad T (2003) *Formations of the Secular: Christianity, Islam, Modernity.* Stanford: Stanford University Press.

Ash J and Simpson P (2016) Geography and Post-Phenomenology. *Progress in Human Geography,* 40(1), 48–66.

Askins K (2015) Being Together: Everyday Geographies and the Quiet Politics of Belonging. *ACME: An International Journal for Critical Geographies*, 14(2), 470–478.

Askins K and Pain R (2011) Contact Zones: Participation, Materiality, and the Messiness of Interaction. *Environment and Planning D*, 29, 803–821.

Atia M (2012) 'A Way to Paradise': Pious Neoliberalism, Islam, and Faith-Based Development. *Annals of the Association of American Geographers*, 102, 808–827.

Atlani L, Caraël M, Brunet J-B, Frasca T, and Chaika N (2000) Social Change and HIV in the Former USSR: The Making of a New Epidemic. *Social Science and Medicine*, 50, 1547–1556.

Audi R (1989) The Separation of Church and State and the Obligations of Citizenship. *Philosophy and Public Affairs*, 18, 259–296.

Audi R (1997) Liberal Democracy and the Place of Religion in Politics. In Audi R and Wolterstorff N (eds) *Religion in the Public Square*. London: Rowman & Littlefield Publishers, pp. 1–66.

Auerbach NN (2012) Delicious Peace Coffee: Marketing Community in Uganda. *Review of Radical Political Economics*, 44(3), 337–357.

Augé M (1988) A Sense for the Other: The Timelessness and Relevance of Anthropology. Stanford, CA: Stanford University Press.

Badiou A (2003) [1997] *Saint Paul: The Foundation of Universalism*. Tran. Ray Brassier. Stanford, CA: Stanford University Press.

Baker C (2009) *Hybrid Church in the City: Third Space Thinking*. London: SCM/Canterbury Press, 2nd Edn.

Baker C (2012) Spiritual Capital and Economies of Grace: Redefining the Relationship between Religion and the Welfare State. *Social Policy and Society*, 11(4), 565–576.

Baker C (2013) Current Themes and Challenges in Urban Theology. *The Expository Times*, 125, 3–12.

Baker C (2016) Faith in the Public Sphere – In Search of a Fair and Compassionate Society for the Twenty-First Century. *Journal of Beliefs and Values*, 37, 259–272.

Baker C (2017a) Ministry and Authenticity. *Anvil: Journal for Theology and Mission*, 33(3), 30–37.

Baker C (2017b) Creating Webs of Connectivity and Hope. *Church Times*. 28 July 2017. Available from www.churchtimes.co.uk/articles/2017/28-july/comment/opinion/creating-webs-of-connectivity-and-hope (last accessed 20/07/18).

Baker C (2018a) Resisting the Transcendent. In Beaumont J (ed.) *The Routledge Handbook of Postsecularity*. London and New York: Routledge.

Baker C (2018b) Postsecularity and a New Urban Politics – Spaces, Places and Imaginaries. In Berking H, Steets S and Schwenk J (eds) *Religious Pluralism and the City: Inquiries into Postsecular Urbanism*. London: Bloomsbury, pp. 81–101.

Baker C and Beaumont J (2011a) Postcolonialism and Religion: New Spaces of 'Belonging and Becoming' in the Postsecular City. In Beaumont J and Baker C (eds) *Postsecular Cities: Space, Theory and Practice*. London: Continuum, pp. 33–49.

Baker C and Beaumont J (2011b) Afterword: Postsecular Cities. In Beaumont J and Baker C (eds) *Postsecular Cities*. Continuum: London, pp. 254–266.

Baker C and Skinner H (2006) *Faith in Action: The Dynamic Connection Between Spiritual and Religious Capital*. Manchester: William Temple Foundation.

Baker J (ed.) (2010) *Curating Worship*. London: SPCK Publishing.

Barbato M (2012) Postsecular Revolution: Religion After the End of History. *Review of International Studies*, 38(5), 1079–1097.

Barbato M and Kratochwil F (2008) Habermas's Notion of a Post-Secular Society. A Perspective from International Relations. EUI Working Papers. MWP 2008/25. Florence, Italy: European University Institute.

Barber M (2017) *Curating Spaces of Hope: A New Definition and Model of Faith-Based Organisations.* William Temple Foundation, Temple Tracts 5(3) Available from https://williamtemplefoundation.org.uk/wp-content/uploads/2017/12/Matthew-Barber-Spaces-of-Hope.pdf (last accessed 02/08/18).

Barbieri W (ed.) (2014) *At the Limits of the Secular: Reflections on Faith and Public Life.* Grand Rapids, MI: Wm. B. Eerdmans Publishing.

Barnes M and Prior D (eds) (2009) *Subversive Citizens: Power, Agency and Resistance in Public Services.* Bristol: Policy Press.

Barnett C (2017) *The Priority of Injustice: Locating Democracy in Critical Theory.* Athens, GA: University of Georgia Press.

Barnett C, Cloke P, Clarke N, and Malpass A (2005) Consuming Ethics: Articulating the Subjects and Spaces of Ethical Consumption. *Antipode*, 37, 23–45.

Barrett A (2017) *Interrupting the Church's Flow: Engaging Graham Ward and Romand Coles in a Radically Receptive Political Theology in the Urban Margins.* Amsterdam: VU University of Amsterdam.

Barthes R (1972) [1957] *Mythologies* trans. Annette Lavers. New York: Hill and Wang.

Bartley J (2006) *Faith and Politics After Christendom: The Church as a Movement for Anarchy.* Milton Keynes: Paternoster Press.

Bartolini N, Chris R, MacKian S, and Pile S (2017) The Place of Spirit: Modernity and the Geographies of Spirituality. *Progress in Human Geography*, 41(3), 338–354.

Bartolini N, MacKian S, and Pile S (eds) (2018) Spaces of Spirituality: An Introduction. In Bartolini N, MacKian S and Pile S (eds) *Spaces of Spirituality*. London and New York: Routledge, pp. 1–24.

Battaglia G (2017) Neo-Hindu Fundamentalism Challenging the Secular and Pluralistic Indian State. *Religions*, 8(10), 216–236.

Bauman Z (1992) *Intimations of Postmodernity.* London: Routledge.

Beaumont J (2008a) Introduction: Faith-Based Organisations and Urban Social Issues. *Urban Studies*, 45, 2011–2017.

Beaumont J (2008b) Faith Action on Urban Social Issues. *Urban Studies*, 45, 2019–2034.

Beaumont J and Baker C (eds) (2011) *Postsecular Cities: Space, Theory and Practice.* London: Continuum Press.

Beaumont J and Cloke P (eds) (2012) *Faith-Based Organisations and Exclusion in European Cities.* Bristol: Policy Press.

Beaumont J and Dias C (2008) Faith-Based Organisations and Urban Social Justice in the Netherlands. *Tijdschrift voor Economishe en Sociale Geografie*, 99, 382–392.

Becker J, Klingan K, Lanz S, and Wildner K (eds) (2013) *Global Prayers: Contemporary Manifestations of the Religious in the City.* Baden: Lars Muller Publishers.

Beckford J (2012) SSSR Presidential Address Public Religions and the Postsecular: Critical Reflections. *Journal for the Scientific Study of Religion*, 51, 1–19.

Beckford R (1998) *Jesus is Dread: Black Theology and Black Culture in Britain.* London: Darton, Longman and Todd.

Bell DM (2001) *Liberation Theology after the End of History: The Refusal to Cease Suffering.* New York: Routledge.

Bender C and Taves A (eds) (2012) *What Matters? Ethnographies of Value in a not so Secular Age.* New York: Columbia University Press.

Bennett B, Dann J, Johnson E, and Reynolds R (eds) (2014) *Once In A Lifetime: City-building After Disaster in Christchurch*. Christchurch, Freerange Press.

Bennett J (2011) *The Enchantment of Modern Life*. Princeton, NJ: Princeton University Press.

Berger P (1967) *The Sacred Canopy*. New York: Doubleday.

Berger P (1969) *A Rumor of Angels: Modern Society and the Rediscovery of the Super-natural*. New York: Doubleday.

Berger P (ed.) (1999) *The Desecularization of the World: Resurgent Religion and World Politics*. Grand Rapids, MI: Eerdmans.

Berger P, Davie G, and Fokas E (2008) *Religious America, Secular Europe?* Aldershot: Ashgate.

Berlant L (2011) *Cruel Optimism*. Durham, NC: Duke University Press.

Berridge V (2005) *Temperance: Its History and Impact on Current and Future Alcohol Policy*. Joseph Rowntree Foundation, York.

Berry D (2017) Religious Strategies of White Nationalism at Charlottesville. *Religion & culture forum*. Available from https://voices.uchicago.edu/religionculture/2017/10/13/religious-strategies-of-white-nationalism-at-charlottesville/ (last accessed 20/07/18).

Bialecki J (2017) Eschatology, Ethics, and Ēthnos: Ressentiment and Christian National-ism in the Anthropology of Christianity. *Religion and Society*, Sept 42–61.

Bielo J (2011) *Emerging Evangelicals – Faith, Modernity and the Desire for Authenticity*. New York: New York University Press.

Bielo J (2013) Belief, Deconversion, and Authenticity Among US Emerging Evangelical. *Ethos*, 40(3), 258–276.

Blaser M (2014) Ontology and Indigeneity: On the Political Ontology of Heterogeneous Assemblages. *Cultural Geographies*, 21(1), 49–58.

Blest P (2017) Over 80,000 People Joined the Biggest-Ever Moral March in North Caro-lina. *The Nation*. 13 February 2017. www.thenation.com/article/over-80000-people-joined-the-biggest-ever-moral-march-in-north-carolina/ (last accessed 20/07/18).

Blok A and Farias I (eds) (2016) *Urban Cosmopolitics: Agencements, Assemblies, Atmo-spheres*. New York: Routledge.

Blond P (1995) Theology and Pluralism. *Modern Theology*, 11, 455–469.

Blond P (1998) Introduction: theology before philosophy. In Blond P (ed.) *Post-secular Philosophy: Between Philosophy and Theology*. London: Routledge, pp. 1–66.

Blond P (2010) *Red Tory*. London: Faber and Faber.

Bloomquist KL (2012) Ekklesia in the Midst of Outrage. *Dialog: A Journal of Theology*, 51(1), 62–70.

Blühdorn I (2006) Self-Experience in the Theme Park of Radical Action?: Social Move-ments and Political Articulation in the Late-Modern Condition. *European Journal of Social Theory*, 9(1), 23–42.

Boeskov K (2017) The Community Music Practice as Cultural Performance: Foundations for Community Music Theory of Social Transformation. *International Journal of Com-munity Music*, 10(1), 85–99.

Bondi L (2013) Between Christianity and Secularity: Counselling and Psychotherapy Provision in Scotland. *Social & Cultural Geography*, 14(6), 668–688.

Bosch D (2011) *Transforming Mission*. Maryknoll, NY: Orbis Books. Twentieth Anni-versary edition.

Bourgois P (2000) Disciplining Addictions: The Bio-politics of Methadone and Heroin in the United States. *Culture Medicine and Psychiatry*, 24, 165–195.

Bourgois P and Hart L (2010) Science, Religion and the Challenges of Substance Abuse Treatment. *Substance Use & Misuse*, 45(14), 2395–2400.

Braidotti R (2008) In Spite of the Times: The Postsecular Turn in Feminism. *Theory, Culture & Society*, 25(6), 1–24.

Braun LN (1997) In from the Cold: Art Therapy with Homeless Men. *Art Therapy – Journal of the American Art Therapy Association*, 14(2), 118–122.

Bretherton L (2010a) *Christianity and Contemporary Politics: The Conditions and Possibilities of Faithful Witness.* Oxford: Wiley-Blackwell.

Bretherton L (2010b) Religion and the Salvation of Urban Politics: Beyond Cooption, Competition and Commodification. In Molendijk A, Beaumont J, and Jedan C (eds) *Exploring the Postsecular: The Religious, the Political and the Urban.* Leiden: Brill, pp. 207–222.

Bretherton L (2011) The Real Battle of St. Paul's Cathedral: The Occupy Movement and Millennial Politics. *Huffington Post.* www.huffingtonpost.co.uk/luke-bretherton/the-real-battle-of-st-pau_b_1065214.html (last accessed 28/08/2015).

Brewin K (2010) *Other: Embracing Difference in a Fractured World.* London: Hodder & Stoughton Ltd.

Brinkman R and Brinkman J (2008) Globalisation and the Nation-State; Dead or Alive. *Journal of Economic Issues*, 42, 425–433.

Brogt E, Grimshaw M, and Baird N (2015) Clergy Views on their Role in City Resilience: Lessons from the Canterbury Earthquakes. *Kotuitui: New Zealand Journal of Social Sciences Online*, 10(2). Available from www.tandfonline.com/doi/full/10.1080/1177083X.1068186 (last accessed 29/03/18).

Bruce S (ed.) (1992) *Religion and Modernisation: Sociologists and Historians Debate the Secularization Thesis.* Oxford: Oxford University Press.

Brueggemann W (1978) *The Prophetic Imagination.* Minneapolis, MN: Fortress Press.

Bryant L, Srnicek N and, Harman G (2011) *The Speculative Turn: Continental Materialism and Realism.* Melbourne: re.press.

Bugyis E (2015) Postsecularism as Colonialism by Other Means. *Critical Research on Religion*, 3(1), 25–40.

Burbridge C (2013) *Faith and the Politics of 'Other': Community Organising Amongst London's Congolese Diaspora.* London: Contextual Theology Centre.

Butler J (1990) *Gender Trouble.* New York: Routledge.

Button M (2005) 'A Monkish Kind of Virtue'? For and Against Humility. *Political Theory*, 33(6), 840–868.

Calhoun C, Juergensmeyer M and VanAntwerpen J (eds) (2011) *Rethinking Secularism.* Oxford: Oxford University Press.

Campbell L and Zimmerman Y (2014) Christian Ethics and Human Trafficking Activism: Progressive Christianity and Social Critique. *Journal of the Society of Christian Ethics*, 34(1), 145–172.

Caputo J (2001) *On Religion.* London: Routledge.

Caputo J (2006) *The Weakness of God: A Theology of the Event.* Bloomington, IN: Indiana University Press.

Carey G (2011) The Occupy Protest at St. Paul's Cathedral – A Parable of our Times. *Daily Telegraph* 27 October.

Carroll A and Norman R (eds) (2016) *Religion and Atheism: Beyond the Divide.* London: Routledge.

Carvalho S (2017) A Week In The Jungle – Volunteering At The Refugee Camp In Calais. *The Huffington Post.* Available from www.huffingtonpost.co.uk/santosh-carvalho/calais-jungle_b_12312682.html (last accessed 21/07/18).

Caryl C (2013) *Strange Rebels: 1979 and the Birth of the 21st Century.* New York: Basic Books.

Casanova J (1994) *Public Religions in the Modern World.* Chicago, IL: University of Chicago Press.

Casanova J (2011) *Public Religions in the Modern World.* Chicago, IL: University of Chicago Press, 2nd Edn.

Castells M (2011) *The Power of Identity.* Oxford: Wiley-Blackwell, 2nd Edn.

Cavanaugh W (1998) *Torture and Eucharist: Theology, Politics, and the Body of Christ.* Oxford: Blackwell.

Cavanaugh W (2014) Invention of Religious-Secular Distinction. In Barbieri WA (ed.) *At the Limits of the Secular: Reflections on Faith and Public Life.* Grand Rapids, MI: Wm. B. Eerdmans, pp. 105–128.

Çavdar A (2013) Negotiations as a Research Methodology. In Becker J, Klingan K, Lanz S and Wildner K (eds) *Global Prayers: Contemporary Manifestations of the Religious in the City.* Baden: Lars Müller Publishers, pp. 198–215.

Caygill H (2002) *Levinas and the Political.* London: Routledge.

Chan SH (2017) Religious Competition and Creative Innovation Amongst Protestant Groups in Hong Kong's Umbrella Movement. *Asian Journal of Religion and Society,* 5(1), 23–48.

Chau A (ed.) (2011) *Religion in Contemporary China: Revitalization and Innovation.* London and New York: Routledge.

Chesters G and Welsh I (2006) *Complexity and Social Movements: Multitudes at the Edge of Chaos.* London: Routledge.

Christianity Uncut Website: http://christianityuncut.wordpress.com (last accessed 27/08/2018).

Christoyannopoulos A (2011) *Christian Anarchism: A Political Commentary on the Gospel.* Imprint Academic: Exeter.

Chu C (ed.) (2014) *Catholicism in China, 1900–Present.* New York: Palgrave Macmillan.

Cistelecan A (2014) The Theological Turn of Contemporary Critical Theory. *Telos,* 167, 8–26.

Citizens UK Website: www.citizensuk.org/ (last accessed 15/05/2018).

Claiborne S (2006) *The Irresistible Revolution: Living as an Ordinary Radical.* Grand Rapids, MI: Zondervan.

Clarke N, Barnett C, Cloke P, and Malpass A (2007) Globalising the Consumer: Doing Politics in an Ethical Register. *Political Geography,* 26, 231–249.

Cloke P (2002) Deliver us from Evil? Prospects for Living Ethically and Acting Politically in Human Geography. *Progress in Human Geography,* 26(5), 587–604.

Cloke P (2010) Theo-ethics and Radical Faith-Based Praxis in the Postsecular City. In Molendijk A, Beaumont J, and Jedan C (eds) *Exploring the Postsecular: The Religious, the Political and the Urban.* Brill, Leiden, pp. 223–241.

Cloke P (2011a) Emerging Postsecular Rapprochement in the Contemporary City. In Beaumont J and Baker C (eds) *Postsecular Cities: Space, Theory and Practice.* London: Continuum, pp. 237–253.

Cloke P (2011b) Emerging Geographies of Evil? Theo-ethics and Postsecular Possibilities. *Cultural Geographies,* 18, 475–493.

Cloke P (2016) Crossover: Working Across Religious and Secular Boundaries. In Cloke P and Pears M (eds) *Mission in Marginal Places: The Theory.* Milton Keynes: Paternoster Press, pp. 145–167.

Cloke P and Beaumont J (2013) Geographies of Postsecular Rapprochement in the City. *Progress in Human Geography*, 37(1), 27–51.

Cloke P and Conradson D (2018) Transitional Organisations, Affective Atmospheres and New Forms of Being-in-common: Post-disaster Recovery in Christchurch, New Zealand. *Transactions of the Institute of British Geographers* (online first). doi. org/10.1111/tran.12240.

Cloke P and Dickinson S (2019) Transitional Ethics and Aesthetics: Re-imagining the Post-disaster City in Christchurch, New Zealand. *Annals of the Association of American Geographers* (in press).

Cloke P and Pears M (eds) (2016a) *Mission in Marginal Places: The Theory.* Milton Keynes: Paternoster Press.

Cloke P and Pears M (eds) (2016b) *Mission in Marginal Places: The Praxis.* Milton Keynes: Paternoster Press.

Cloke P and Sutherland C (Forthcoming) Conceptualising Social Protection: Evaluating Third Sector Practice between the Universal and the Local.

Cloke P and Williams A (2018) Geographical Landscapes of Religion. In Baker C, Crisp B, and Dinham A (eds) *Re-imagining Religion and Belief for 21st Century Policy and Practice.* Cambridge: Policy Press, pp. 33–54.

Cloke P, Beaumont J, and Williams A (eds) (2013) *Working Faith: Faith-based Organisations and Urban Social Justice.* Milton Keynes: Paternoster Press.

Cloke P, Dickinson S, and Tupper S (2017) The Christchurch Earthquakes 2010, 2011: Geographies of an Event. *New Zealand Geographer*, 73(2), 68–80.

Cloke P, Johnsen S, and May J (2005) Exploring Ethos? Discourses of 'Charity' in the Provision of Emergency Services for Homeless People. *Environment and Planning A*, 37, 385–402.

Cloke P, Johnsen S, and May J (2007) The Periphery of Care: Emergency Services for Homeless People in Rural Areas. *Journal of Rural Studies*, 23, 387–401.

Cloke P, May J, and Johnsen S (2007) Ethical Citizenship? Volunteers and the Ethics of Providing Services for Homeless People. *Geoforum*, 38, 1089–1101.

Cloke P, May J, and Johnsen S (2010) *Swept Up Lives? Re-envisioning the Homeless City.* Oxford: Wiley-Blackwell.

Cloke P, May J, and Williams A (2017) The geographies of food banks in the meantime. *Progress in Human Geography,* 41(6), 703–726.

Cloke P, Sutherland C, and Williams A (2016) Postsecularity, Political Resistance, and Protest in the Occupy Movement. *Antipode*, 48(3), 497–523.

Cloke P, Thomas S, and Williams A (2012) Radical Faith Praxis? Exploring the Changing Theological Landscape of Christian Faith Motivation. In Beaumont J and Cloke P (eds) *Faith-based Organisations and Exclusion in European Cities.* Bristol: Policy Press, pp. 105–126.

Cloke P, Thomas S, and Williams A (2013a) Faith in Action: Faith-based Organization, Welfare and Politics in the Contemporary City. In Cloke P, Beaumont J, and Williams A (eds) *Working Faith: Faith-based Organisations and Urban Social Justice.* Milton Keynes: Paternoster, pp. 1–24.

Cloke P, Williams A, and Thomas S (2013b) CAP in Two Guises: A Comparison of Christians Against Poverty and Church Action on Poverty. In Cloke P, Beaumont J, and Williams A (eds) *Working Faith: Faith-based Organisations and Urban Social Justice.* Milton Keynes: Paternoster, pp. 25–46.

Coles R (1997) *Rethinking Generosity: Critical Theory and the Politics of Caritas.* Ithaca, NY: Cornell University Press.

Coles R (2001) Traditio: Feminists of Color and the Torn Virtues of Democratic Engagement. *Political Theory*, 29(4), 488–516.

Connolly W (1999) *Why I am Not a Secularist.* Minneapolis, MN: University of Minnesota Press.

Connolly W (2005) The Evangelical-Capitalist Resonance Machine. *Political Theory*, 33, 869–886.

Connolly W (2008) *Capitalism and Christianity, American Style.* Durham, NC: Duke University Press.

Conradson D (2003) Spaces of Care in the City: The Place of a Community Drop-in Centre. *Social and Cultural Geography*, 4, 507–525.

Conradson D (2013) Somewhere Between Religion and Spirituality? Places of Retreat in Contemporary Britain. In Hopkins P, Kong L, and Olson E (eds) *Religion and Place: Landscape, Politics, and Piety*. New York, NY: Springer, pp. 185–202.

Cooke M (2006) *Re-presenting the Good Society.* Cambridge, MA: MIT Press.

Cox J (2015) Taxpayers Subsidise Big Business by an Estimated £11 Billion a Year. Citizens UK. 12 April 2015. www.citizensuk.org/taxpayer (last accessed 20/07/18).

Cramer D, Howell J, Martens P, and Tran J (2014) Theology and Misconduct: The Case of John Howard Yoder. *The Christian Century*, 131(17), 20–24.

Critchley S (2012) *The Faith of the Faithless: Experiments in Political Theology*. London: Verso Books.

Crouch W (2011) *The Strange Non-Death of Neo-Liberalism*. Cambridge: Polity Press.

Cupples J and Glynn K (2009) Countercartographies: (New Zealand) Cultural Studies/ Geographies of the City. *New Zealand Geographer*, 68, 1–5.

Cvetkovich A (2012) *Depression: A Public Feeling.* Durham, NC: Duke University Press.

Dabashi H (2012) *The Arab Spring.* London: Zed.

Dalferth I (2010) Post-secular Society: Christianity and the Dialectics of the Secular. *Journal of the American Academy of Religion*, 78, 317–345.

Daniel W and Marsh C (2008) Russia's 1997 Law on Freedom of Conscience in Context and Retrospect. In Daniel W, Berger P, and Marsh C (eds) *Perspectives on Church–State Relations in Russia.* Waco TX: Baylor University Press.

Darling J (2010) A City of Sanctuary: The Relational Re-imagining of Sheffield's Asylum Politics. *Transactions of the Institute of British Geographers*, 35, 125–140.

Davelaar M, Williams A, and Beaumont J (2013) Adventures at a Border-Crossing: The Society for Diaconal Social Work in Rotterdam, the Netherlands. In P Cloke, J Beaumont, and A Williams (eds), *Working Faith: Faith-based Organizations and Urban Social Justice.* Milton Keynes: Paternoster Press, pp. 165–184.

Davie G (2007) *The Sociology of Religion.* London: SAGE.

Davie G (2015) *Religion in Britain: A Persistent Paradox.* Oxford: Wiley-Blackwell.

Davies T, Isakjee A, and Dhesi S (2017) Violent Inaction: The Necropolitical Experience of Refugees in Europe. *Antipode*, 49(5), 1263–1284.

Davies W (2011) The Political-economy of Unhappiness. *New Left Review*, 71, 65–80.

Davis C, Milbank J, and Žižek S (2011) *The Monstrosity of Christ: Paradox or Dialectic?* Cambridge, MA: MIT Press.

Davis J (1997) Building from the Scraps: Art Therapy within a Homeless Community. *Art Therapy – Journal of the American Art Therapy Association*, 14(3), 210–213.

Davis M (1990) *City of Quartz: Excavating the Future in Los Angeles.* London: Verso.

Davis M (2007) *Planet of Slums.* London: Verso.

Dawkins R (2007) *The God Delusion.* London: Bantam.

Day D (1997) *The Long Loneliness*. New York, NY: HarperCollins.

de Vries H (2006) Introduction: Before, Around and Beyond the Theologico-political. In de Vries H and Sullivan L (eds) *Political Theologies: Public Religions in a Postsecular World*. New York: Fordham University Press, pp. 1–88.

Deleuze G and Guattari F (1994) *What is Philosophy?* New York, NY: Columbia University Press.

della Dora V (2016) Infrasecular Geographies: Making, Unmaking and Remaking Sacred Space. *Progress in Human Geography*, 42(1), 44–71.

Della Porta D (2005) Multiple Belongings, Tolerant Identities, and the Construction of 'Another Politics': Between the European Social Forum and the Local Social Fora. In Della Porta D and Tarrow S (eds) *Transnational Protest and Global Activism*. Oxford: Rowman & Littlefield Publishers, pp. 175–202.

Democracy Now! (2017) Cornel West & Rev. Traci Blackmon: Clergy in Charlottesville Were Trapped by Torch-Wielding Nazis. *Democracy Now!* 14 August 2017. Available from www.democracynow.org/2017/8/14/cornel_west_rev_toni_blackmon_clergy (last accessed 20/07/18).

Deneulin S and Rakodi C (2011) Revisiting Religion: Development Studies Thirty Years On. *World Development*, 39(1), 45–54.

Depoortere F (2008) *Christ in Postmodern Philosophy: Gianni Vattimo, Rene Girard, and Slavoj Žižek*. London: T&T Clark International.

Derrida J (1994) *Specters of Marx: The State of the Debt, the Work of Mourning, and the New International*, trans. Peggy Kamuf. London: Routledge.

Derrida J (2002) 'Faith' and 'Knowledge': The Two Sources of 'Religion' at the Limits of Reason Alone. In Derrida J and Vattimo G (eds) *Religion*. Cambridge: Polity Press, pp. 1–78.

Dewsbury JD (2000) Performativity and the Event: Enacting a Philosophy of Difference. *Environment and Planning D*, 18, 473–496.

Dewsbury JD (2003) Witnessing Space: 'Knowledge without Contemplation'. *Environment and Planning A*, 35(11), 1907–1932.

Dewsbury JD (2007) Unthinking Subjects: Alain Badiou and the Event of Thought in Thinking Politics. *Transactions of the Institute of British Geographers*, 32(4), 443–459.

Dewsbury JD and Cloke P (2009) Spiritual Landscapes: Existence, Performance and Immanence. *Social & Cultural Geography*, 10, 695–711.

Dillon M (2010) Can Post-Secular Society Tolerate Religious Differences? *Sociology of Religion*, 71, 139–156.

Dinham A and Lowndes V (2008) Religion, Resources, and Representation: Three Narratives of Faith Engagement in British Urban Governance. *Urban Affairs Review*, 43(6), 817–845.

Dionne E Jnr (2008) *Souled Out: Reclaiming Faith and Politics After the Religious Right*. Princeton, NJ: Princeton University Press.

Diprose G (2015) Negotiating Contradiction: Work, Redundancy and Participatory Art. *Area*, 47(3), 246–253.

Dittmer J (2008) The Geographical Pivot of (the End of) History: Evangelical Geopolitical Imaginations and Audience Interpretation of Left Behind. *Political Geography*, 27(3), 280–300.

Dittmer J and Sturm T (2010) *Mapping the End Times: American Evangelical Geopolitics and Apocalyptic Visions*. Farnham: Ashgate Publishing.

Doak M (2007) The Politics of Radical Orthodoxy: A Catholic Critique. *Theological Studies*, 68(2), 368–393.

Doel M (1994) Deconstruction on the Move: From Libidinal Economy to Liminal Materialism. *Environment and Planning A*, 26, 1041–1059.

Doel M (1995) Bodies without Organs: Schizoanalysis and Deconstruction. In Pile S and Thrift N (eds) *Mapping the Subject: Geographies of Cultural Transformation.* London: Routledge.

Dossett W (2013) Addiction, Spirituality and 12-Step Programmes. *International Social Work*, 56(3), 369–383.

Douch P (2005) *The Busker's Guide to Inclusion.* Eastleigh: Common Threads Publications Ltd.

Duff C (2017) The Affective Right to the City. *Transactions of the Institute of British Geographers*, 42(4), 516–529.

Dunnington K (2011) *Addiction and Virtue.* Downers Grove, IL: InterVarsity Press USA.

Dwyer C (2016) Why Does Religion Matter for Cultural Geographers? *Social & Cultural Geography*, 17, 758–762.

Dwyer C and Parutis V (2013) 'Faith in the system?' State-funded Faith Schools in England and the Contested Parameters of Community Cohesion. *Transactions of the Institute of British Geographers*, 38, 267–284.

Eagleton T (2009) *Trouble with Strangers: A Study of Ethics.* Chichester: John Wiley & Sons Ltd.

Eagleton T (2010) *Reason, Faith and Revolution.* London: Yale University Press.

Eagleton T (2011) *On Evil.* New Haven, CT: Yale University Press.

Eberle C (2002) *Religious Conviction in Liberal Politics.* Cambridge: Cambridge University Press.

Eder K (2002) Europaische sakularisierung- ein sonderweg in die postsakulare gesellschaft. *Berliner Journal für Soziologie*, 3, 331–343.

Eder K (2006) Post-Secularism: A Return to the Public Sphere. *Eurozine.* 17 August.

Ehrkamp P and Nagel C (2014) 'Under the Radar': Undocumented Migrants, Christian Faith Communities, and the Precarious Spaces of Welcome in the US South. *Annals of the Association of American of Geographers*, 104, 319–328.

Eisenstein C (2011) *Sacred Economics: Money, Gift, and Society in the Age of Transition.* New York, NY: Evolver Editions.

Ekklesia (2011) 'Sermon on the Steps' Response to the Bishop of London's Ultimatum. Available from www.ekklesia.co.uk/node/15616 (last accessed 28/08/2015).

End Hunger UK (2017) End Hunger UK: A Menu to End Hunger in the UK. Available from http://endhungeruk.org/wp-content/uploads/2017/12/A-Menu-to-End-Hunger-in-the-UK.pdf.

Epstein B (2002) The Politics of Prefigurative Community: The Non-Violent Direct Action Movement. In Duncombe S (ed.) *Cultural Resistance Reader.* London: Verso, pp. 333–346.

Everett A (2018) *After the Fire: Finding Words for Grenfell.* London: Canterbury Press Norwich.

Fairbanks R (2009) *How it Works: Recovering Citizens in Post-welfare Philadelphia.* Chicago: University of Chicago.

Faith in the City (1985) *Faith in the City: The Report of the Archbishop of Canterbury's Commission on Urban Priority Areas.* London: Church House Publishing.

Farias I and Bloc A (2016) Introducing Urban Cosmopolitics: Multiplicity and the Search for a Common World. In Bloc A and Farias I (eds) *Urban Cosmopolitics: Agencements, Assemblies, Atmospheres.* London: Routledge, pp. 1–22.

Featherstone D, Ince A, Mackinnon D, Strauss K, and Cumbers A (2012) Progressive Localism and the Construction of Political Alternatives. *Transactions of the Institute of British Geographers*, 37(2), 177–182.

Ferguson H (1995) *Melancholy and the Critique of Modernity Søren Kierkegaard's Religious Psychology.* London: Routledge.

Ferguson J and Gupta A (2002) Spatialising States: Toward an Ethnography of Neoliberal Governmentality. *American Ethnologist*, 29(4), 981–1002.

Finley S (2000) 'Dream Child': The Role of Poetic Dialogue in Homeless Research. *Qualitative Inquiry*, 6(3), 432–434.

Finley S and Finley M (1999) Sp'ange: A Research Story. *Qualitative Inquiry*, 5(3), 313–337.

Firth R (2016) Somatic Pedagogies: Critiquing and Resisting the Affective Discourse of the Neoliberal State from an Embodied Anarchist Perspective. *Ephemera*, 16(4), 121–142.

Fisher A (2017) *Big Hunger: The Unholy Alliance between Corporate America and Anti-Hunger Groups.* Cambridge, MA: The Massachusetts Institute of Technology Press.

Fisher AS (2005) Developing an Ethics of Practice Applied Theatre: Badiou and Fidelity to the Truth of the Event. *Research in Drama Education: The Journal of Applied Theatre and Performance*, 10(2), 247–252.

Fisher M (2009) *Capitalist Realism: Is There No Alternative?* Winchester: Zed Books.

Foucault M (1980) *Power/Knowledge: Selected Interviews and Other Writings, 1972–1977.* Gordon C (ed.), trans. Marshall L, Mepham J, and Soper K. New York: Pantheon Books.

Foucault M (2005) *The Hermeneutics of the Subject: Lectures at the Collège de France 1981–1982*, trans. Burchell G. New York, NY: Picador.

Franks P and McAloon J (2016) *Labour: The New Zealand Labour Party 1916–2016.* Wellington: Victoria University Press.

Fraser S and Valentine K (2008) *Substance and Substitution: Methadone Subjects in Liberal Societies.* New York: Palgrave Macmillan.

Frisk L (2011) The Practice of Mindfulness: From Buddhist Practice to Secular Mainstream in a Postsecular Society. Paper given in Religion in Postsecular Society Åbo, Finland, 15–17 June.

Furness S and Gilligan P (2010) *Religion, Belief and Social Work.* Bristol: Policy Press.

Fuste-Forne F (2017) Building Experiencescapes in Christchurch. *Landscape Review*, 17, 44–57.

Galindez S (2012) Black Churches to Energize Occupy. *The Indypendent.* Available from www.indypendent.org/2012/01/09/black-churches-energize-occupy (last accessed 28/08/2015).

Ganiel G (2006) Emerging from Evangelical Subculture in Northern Ireland: An Analysis of Zero28 and Ikon Community. *International Journal for the Study of the Christian Church*, 6(1), 38–48.

Ganiel G and Marti G (2014) Northern Ireland, American and the Emerging Church Movement: Exploring the Significance of Peter Rollins and the Ikon Collective. *Journal of the Irish Society for the Academic Study of Religions*, 1(1), 26–47.

Gao J (2013) Deleuze's Concept of Desire. *Deleuze Studies*, 7(3), 406–420.

Gao Q, Qian J, and Yuan Z (2018) Multi-scaled Secularization or Postsecular present? Christianity and Migrant Workers in Shenzhen, China. *Cultural Geographies*, 25(4), 553–570.

Garbin D and Strhan A (2017) *Religion and the Global City.* London: Bloomsbury Publishing.

Garrigan S (2010) *The Real Peace Process: Worship, Politics and the End of Sectarianism.* London: Equinox.

Garthwaite K (2016) *Hunger Pains: Life Inside Foodbank Britain.* Bristol: Policy Press.

Gibson-Graham JK (2006) *A Postcapitalist Politics.* Minneapolis, MN: University of Minnesota Press.

Gilbert J (2017) What is Acid Corbynism? *Red Pepper.* Available from www.redpepper. org.uk/what-is-acid-corbynism/ (last accessed 28/07/2018).

Gökarıksel B and Secor A (2014) The Veil, Desire, and the Gaze: Turning the Inside Out. *Signs: Journal of Women in Culture and Society*, 40(1), 177–200.

Gökarıksel B and Secor A (2015) Post-secular Geographies and the Problem of Pluralism: Religion and Everyday Life in Istanbul, Turkey. *Political Geography*, 46, 21–30.

Göle N (2010) The Civilizational, Spatial and Sexual Powers of the Secular. In Warner M, Vantantwerpen J, and Calhoun C (eds) *Varieties of Secularism in a Secular Age.* Cambridge, MA: Harvard University Press, pp. 243–264.

Gordon P (2011) What Hope Remains? *The New Republic*, 14 December. Available from https://newrepublic.com/article/98567/jurgen-habermas-religion-philosophy (last accessed 6 June 2017).

Gorski P and Altınordu A (2008) After Secularization? *Annual Review of Sociology*, 34, 55–85.

Gorski P, Kim D, Torpey J, and Van Antwerpen J (eds) (2012) *The Post-Secular in Question.* New York: New York University Press.

Graham E (2013) *Between a Rock and a Hard Place: Public Theology in a Postsecular Age.* London: SCM Press.

Graham E (2018) How to Speak of God? Toward a Postsecular Apologetics. *Practical Theology*, 11(3), 206–217.

Graham E and Lowe S (2009) *What Makes a Good City? Public Theology and the Urban Church.* London: Darton, Longman and Todd.

Gray J (2010) Red Tory, By Phillip Blond. *Independent.* 2 April 2010. Available from www.independent.co.uk/arts-entertainment/books/reviews/red-tory-by-phillip-blond-1933475.html (last accessed 08/06/18).

Greed C (1994) *Women and Planning: Creating Gendered Realities.* London: Routledge.

Greed C (2011) A Feminist Critique of the Postsecular City. In Beaumont J and Baker C (eds) *Postsecular Cities: Space Theory and Practice.* London and New York: Continuum, pp. 104–119.

Greeley AM (1966) After Secularity: The Neo-Gemeinschaft Society – A Post-Christian Postscript. *Sociological Analysis*, 27(3), 119–127.

Green E (2017) Evangelicals Are Bitterly Split Over Advising Trump. *The Atlantic.* 22 August 2017. Available from www.theatlantic.com/politics/archive/2017/08/evangelical-advisers-trump/537513/ (last accessed 20/07/18).

Greenstreet W (2006) Past and Present Discourses. In Greenstreet W (ed.) *Integrating Spirituality in Health and Social Care.* Oxford: Radcliffe, pp. 20–31.

Gutiérrez G (1988) *A Theology of Liberation: History, Politics, and Salvation.* London: SCM Press.

Habermas J (1995) Reconciliation Through the Public Use of Reason: Remarks on John Rawls's Political Liberalism. *The Journal of Philosophy*, 92, 109–131.

Habermas J (2002) *Religion and Rationality: Essays on Reason, God and Modernity.* Cambridge, MA: MIT Press.

Habermas J (2005) Equal Treatment of Cultures and the Limits of Postmodern Liberalism. *Journal of Political Philosophy*, 13, 1–28.

Habermas J (2006a) Religion in the Public Sphere. *European Journal of Philosophy*, 14(1), 1–25.

Habermas J (2006b) *Time of Transitions*, trans. Schott G. Cambridge: Polity Press.

Habermas J (2006c) Pre-political Foundations of the Democratic Constitutional State? In Habermas J and Ratzinger J (eds) *The Dialectics of Secularization: On Reason and Religion*, trans. McNeil B. San Francisco: Ignatius, pp. 19–52.

Habermas J (2008) Notes on a Post-secular Society. Signandsight.com. 18 June 2008. 1–23. Available from www.signandsight.com/features/1714.html (last accessed 08/07/18).

Habermas J (2010) An Awareness of What is Missing. In Habermas J, Brieskorn N, Reder M, Ricken F, and Schmidt J (eds) *An Awareness of What Is Missing: Faith and Reason in a Postsecular Age*. Cambridge: Polity Press, pp. 15–23.

Habermas J (2010) *An Awareness of What is Missing: Faith and Reason in a Post-secular Age*. Cambridge: Polity Press.

Habermas J (2013) Reply to my Critics. In Calhoun C, Mendieta E, and VanAntwerpen J (eds) *Habermas and Religion*. Cambridge: Polity Press, pp. 347–390.

Habermas J and Ratzinger J (2006) *The Dialectics of Secularization: On Reason and Religion*. San Francisco, CA: Ignatius Press.

Hackworth J (2010) Compassionate Neoliberalism? Evangelical Christianity, the Welfare State, and the Politics of the Right. *Studies in Political Economy*, 86, 83–108.

Hackworth J (2012) *Faith-Based: Religious Neoliberalism and the Politics of Welfare*. Athens, GA: University of Georgia Press.

Häkli J and Kallio KP (2014) Subject, Action, and Polis: Theorizing Political Agency. *Progress in Human Geography*, 38(2), 181–200.

Hammond K and Richey J (eds) (2015) *The Sage Returns: Confucian Revival in Contemporary China*. Albany, NY: State University of New York.

Haraway D (2016) *Staying With the Trouble: Making Kin in the Chthulucene*. Durham, NC: Duke University Press.

Hardt M and Negri A (2001) *Empire*. Cambridge, MA: Harvard University Press.

Harrington A (2007) Habermas and the 'Post-Secular Society'. *European Journal of Social Theory*, 10(4), 543–560.

Harrison H (2011) Global Modernity, Local Community, and Spiritual Power in the Shanxi Catholic Church. In Chau A (ed.) *Religion in Contemporary China: Revitalization and Innovation*. London and New York: Routledge.

Harvey J (2011) Ecumenical Action in the Gorbals. *Theology in Scotland*, 13(2), 57–64.

Hastings A, Bailey N, Bramley G, and Gannon M (2017) Austerity Urbanism in England: The 'Regressive Redistribution' of Local Government Services and the Impact on the Poor and Marginalised. *Environment and Planning A*, 49(9), 2007–2024.

Hauerwas S (1983) *The Peaceable Kingdom*. Chicago, IL: University of Notre Dame Press.

Hauerwas S (2000) *A Better Hope*. Ada, MI: Brazos Press.

Hauerwas S and Coles R (2008) *Christianity, Democracy and the Radical Ordinary*. Eugene, OR: Cascade Books.

Hauerwas S, Wells S, Bretherton L., and Rook R (2010) *Living Out Loud: Conversations About Virtue, Ethics and Evangelicalism*. Milton Keyes: Paternoster.

Hayward B (2012) Canterbury's Political Quake. Available from www.stuff.co.nz/the-press/opinion/perspective/6664104/Canterburys-political-quake/ (last accessed 09/08/16).

Healy S (2013) Affective Dissent. *Cosmopolitan Civil Societies Journal*, 5(2), 114–130.

Heelas P and Woodhead L (2005) *The Spiritual Revolution: Why Religion is Giving Way to Spirituality*. Oxford: Blackwell.

Hemming P (2011) Meaningful Encounters? Religion and Social Cohesion in the English Primary School. *Social & Cultural Geography*, 12(1), 63–81.

Herman A, Beaumont J, Cloke P, and Walliser A (2012) Spaces of Postsecular Engagement in Cities. In Beaumont J and Cloke P (eds) *Faith-based Organisations and Exclusion in European Cities.* Bristol: Policy Press, pp. 59–80.

Hibbard S (2018) 'FGM has Stopped in Wales' but Women still 'Persecuted'. BBC News, 14 June 2018. Available from www.bbc.co.uk/news/uk-wales-44440167 (last accessed 30/10/18).

Hilgers M (2013) Embodying Neoliberalism: Thoughts and Responses to Critics. *Social Anthropology*, 21(1), 75–89.

Hitchins C (2008) *God is not Great.* London: Atlantic.

HM Treasury (2015) *Spending Review and Autumn Statement 2015.* London: HMSO. Available from www.gov.uk/government/uploads/system/uploads/attachment_data/file/479749/52229_Blue_Book_PU1865_Web_Accessible.pdf (last accessed 20/06/18).

Holland J (2015) *Peter Maurin's Ecological Lay New Monasticism: A Catholic Green Revolution Developing Rural Ecovillages, Urban Houses of Hospitality, & Eco-Universities for a New Civilization.* Washington, DC: Pacem in Terris Press.

Holloway J (2003) Make-Believe: Spiritual Practice, Embodiment and Sacred Space. *Environment and Planning A*, 35(11), 1961–1974.

Holloway J (2011) Tracing the Emergent in the Geographies of Religion and Belief. In Bailey A, Brace C, Carter S, Harvey D, Hill J, and Thomas N (eds) *Emerging Geographies of Belief.* Newcastle: Cambridge Scholars, pp. 30–53.

Holloway J (2011) Spiritual Life. In Del Casino VJ, Thomas ME, Cloke P, and Panelli R (eds) *A Companion to Social Geography.* Chichester: Blackwell Publishing Ltd, pp. 363–384.

Holloway J (2013) The Space that Faith Makes: Towards a (Hopeful) Ethos of Engagement. In Hopkins P, Kong L, and Olson E (eds) *Religion and Place: Landscape, Politics and Piety.* Dordrecht: Springer Publishing, pp. 203–218.

Holloway J and Valins O (2010) Editorial: Placing Religion and Spirituality in Geography. *Social & Cultural Geography*, 3, 5–9.

Holton R (2011) *Globalisation and the Nation-State.* Basingstoke: Palgrave Macmillan, 2nd Edn.

Hooks B (1982) *Ain't I a Woman?: Black Women and Feminism.* London: Pluto Press.

Hopkins P (2017) Social Geography 1: Intersectionality. *Progress in Human Geography* doi.org/10.1177/0309132517743677.

Hopkins P, Kong L, and Olson E (eds) (2013) *Religion and Place: Landscape, Politics and Piety.* Dordrecht: Springer Publishing.

Horton J and Kraftl P (2009) Small Acts, Kind Words and 'Not Too Much Fuss': Implicit Activisms. *Emotion, Space and Society*, 2(1), 14–23, 1755–4586.

Housing Justice Homelessness Sunday Page: www.housingjustice.org.uk/Event/homeless-Sunday-2018. (last accessed 16/05/16).

Howson C (2011) *A Just Church: 21st Century Liberation Theology in Action.* London: Continuum International Publishing Group.

Huffschmid A (2013) From Padre Mugica to Santa Muerte? Liberation Spirits and Religious Mutations in Urban Space in Latin America. In Becker J, Klingan K, Lanz S, and Wildner K (eds) *Global Prayers: Contemporary Manifestations of the Religious in the City.* Baden: Lars Müller Publishers, pp. 392–407.

Inbar Y, Pizzaro D, and Bloom P (2011) Conservatives are More Easily Disgusted than Liberals. *Cognition and Emotion*, 23, 714–725.

Isin EF and Rygiel K (2007) Abject Extrality: Frontiers, Zones and Camps. In Dauphinee E and Masters C (eds) *The Logics of Biopower and the War on Terror: Living, Dying, Surviving.* Basingstoke: Palgrave Macmillan, pp. 181–203.

Ivinson G and Renold E (2013) Subjectivity, Affect and Place: Thinking with Deleuze and Guattari's Body without Organs (BwO) to Explore a Young Girl's Becomings in a Post-industrial Locale. *Subjectivity*, 6(4), 369–390.

Jabir T (2015) *Secularism, 'Indian secularism' and Post-secular Discourses in India.* Available from www.cedl.ac.in/download.php?id=25 (last accessed 25/05/18).

Jaffrelot C (1999) *The Hindu National Movement and Indian Politics: 1925 to the 1990s. With New Afterword.* New Delhi: Penguin Books.

James M (2014) *A Brief History of Seven Killings.* London: Oneworld Publications.

Jamoul L and Wills J (2008) Faith in Politics. *Urban Studies*, 45(10), 2035–2056.

Jedan, C (2010) Beyond the Secular? Public Reason and the Search for a Concept of Postsecular Legitimacy. In Molendijk A, Beaumont J, and Jedan C (eds) *Exploring the Postsecular: The Religious, the Political and the Urban.* Leiden, the Netherlands: Brill, pp. 311–327.

Jenkins J (2017) Meet the Clergy who Stared Down White Supremacists in Charlottesville. *Think Progress.* Available from https://thinkprogress.org/clergy-in-charlottesville-e95752415c3e/ (last accessed 20/07/18).

Jensen D (2006) *Endgame: Vol. II, Resistance.* New York: Seven Stories Press.

Jensen T (2011) On the Emotional Terrain of Neoliberalism. *Journal of Aesthetics and Protest*, 8. Available from http://joaap.org/issue8/jensen.htm (last accessed 07/07/18).

Ji Z (2006) Non-Institutional Religious Re-Composition among the Chinese Youth. *Social Compass*, 53, 535–549.

Joas H (2008) *Do We Need Religion? On the Experience of Self-transcendence*, trans. Skinner A. Boulder, CO: Paradigm Publishers.

Johansson M and Kociatkiewicz J (2011) City Festivals: Creativity and Control in Staged Urban Experiences. *European Urban and Regional Studies*, 18, 392–405.

Jones R (2008) *Progressive & Religious: How Christian, Jewish, Muslim, and Buddhist Leaders are Moving Beyond Partisan Politics and Transforming American Public Life.* Michigan: Rowman & Littlefield Publishers.

Jones R and Heley J (2016) Post-pastoral? Rethinking Religion and the Reconstruction of Rural Space. *Journal of Rural Studies*, 45, 15–23.

Judson Memorial Church Homepage: www.judson.org/index.php (last accessed 30/09/16).

Juergensmeyer M, Griego D, and Soboslai J (2015) *God in the Tumult of the Global Square.* Oakland, CA: University of California Press.

Juris JS (2005) Social Forums and the Margins: Networking Logics and the Cultural Politics of Autonomous Spaces. *Ephemera*, 5(2), 253–272.

Karpov V (2010) Desecularisation: A Conceptual Framework. *Journal of Church and State*, 52, 232–270.

Karpov V (2013) The Social Dynamics of Russia's Desecularisation: A Comparative and Theoretical Perspective. *Religion, State and Society*, 41, 254–283.

Kaufmann (2009) Locating the Postsecular. *Religion & Literature*, 41(3), 68–73.

Kearney R (2011) *Anatheism: Returning to God After God.* New York: Columbia University Press.

Kershaw B (1992) *The Politics of Performance: Radical Theatre as Cultural Intervention.* London: Routledge.

Kessler E and Arkush M (2008) *Keeping Faith in Development: The Significance of Interfaith Relations in the Work of Humanitarian Aid and International Development Organisations.* Cambridge: Woolf Institute.

Khanum F (2012) A Common Agenda: Interfaith Dialogue and Faith Secular Partnership. Paper in Africa–UK Annual Conference 2012: Faith and Development in Africa. 3 December 2012. Chelsea Old Town Hall, King's Road, London.

Kidd SA (2009) 'A Lot of Us Look at Life Differently': Homeless Youths and Art on the Outside. *Cultural Studies–Critical Methodologies*, 9(2), 345–367.

Kimberly J, Troy S, Glover D, and Parry DC (2004) Leisure Spaces as Potential Sites for Interracial Interaction: Community Gardens in Urban Areas. *Journal of Leisure Research*, 36(3), 336–355.

King N (2016) *No Borders – The Politics of Immigration Control and Resistance.* London: Zed Books.

Klein N (2017) *No Is Not Enough: Defeating the New Shock Politics.* London: Allen Lane.

Kong L (2010) Global Shifts, Theoretical Shifts: Changing Geographies of Religion. *Progress in Human Geography*, 34(6), 755–776.

Kuhn R (1976) *The Demon of Noontide: Ennui in Western Literature.* Princeton, NJ: Princeton University Press.

Kumar P (2008) *Limiting Secularism: The Ethics of Co-existence in Indian Literature and Film.* Minneapolis: Minnesota University Press.

Kumm B (2013) Finding Healing through Songwriting: A Song for Nicolette. *International Journal of Community Music*, 6(2), 205–217.

Kuznetsova I and Round J (2014) Communities and Social Care in Russia: The Role of Muslim Welfare Provision in Everyday Life in Russia's Tartarstan Region. *International Social Work*, 57, 486–496.

Lafonte C (2007) Religion in the Public Sphere: Remarks on Habermas's Conception of Public Deliberation in Postsecular Societies. *Constellations*, 14, 239–259.

Lambie-Mumford H (2017) *Hungry Britain: The Rise of Food Charity.* Bristol: Policy Press.

Lancione M (2014) Entanglements of Faith: Discourses, Practices of Care and Homeless People in an Italian City of Saints. *Urban Studies*, 51, 3062–3078.

Landy J and Saler M (eds) (2009) *The Re-enchantment of the World.* Stanford, CA: Stanford University Press.

Lane BC (2002) *Landscapes of the Sacred: Geography and Narrative in American Spirituality.* Baltimore, MD: The Johns Hopkins University Press.

Larner W (2000) Neo-liberalism Policy, Ideology, Governmentality. *Studies in Political Economy*, 63(1), 5–25.

Latour B (1993) *We Have Never Been Modern*, trans. C Porter. Cambridge, MA: Harvard University Press.

Latour B (2005) *Reassembling the Social.* Oxford: Oxford University Press.

Lawson V (2007) Geographies of Care and Responsibility. *Annals of the Association of American Geographers*, 97, 1–11.

Lawson V and Elwood S (2013) Encountering Poverty: Space, Class, and Poverty Politics. *Antipode*, 46(1), 209–228.

Lee L (2015) *Recognizing the Non-religious: Reimagining the Secular.* Oxford: Oxford University Press.

Lee L (2017) Godlessness in the Global City. In Garbin D and Strhan A (eds) *Religion and the Global City.* London: Bloomsbury Publishing, pp. 135–152.

Levinas E (1978) *Existence and Existents*, trans. Lingis A. The Hague: M. Nijhoff.

Levine G (2008) *Darwin Loves You: Natural Selection and the Re-enchantment of the World.* Princeton, NJ: Princeton University Press.

Lewicki A and O'Toole T (2016) Acts and Practices of Citizenship: Muslim Women's Activism in the UK. *Ethnic and Racial Studies*, 40, 152–171.

Lewis T (2016) Spirited Publics? Post-secularism, Enchantment and Enterprise on Indian Television. In Marshall P, D'Cruz G, McDonald S, and Lee K. (eds) *Contemporary Publics*. London: Palgrave Macmillan.

Ley D (2011) Preface: Towards a Postsecular City? In Beaumont J and Baker C (eds) *Postsecular Cities: Space, Theory and Practice*. London: Continuum, pp. xii–xiv.

Ley D and Tse J (2013) *Homo Religiosus*? Religion and Immigrant Subjectivities. In Hopkins P, Kong L, and Olson E (eds) *Religion and Place: Landscape, Politics and Piety*. New York: Springer, pp. 149–166.

Lidman L (2014) In Uganda, Coffee Co-op Blends Jewish, Muslim and Christian Farmers. *The Times of Israel*. Available from www.timesofisrael.com/in-uganda-coffee-co-op-blends-jewish-muslim-and-christian-farmers/ (last accessed 27/07/18).

Lineham P (2017) *Sunday Best: How Religion Shaped New Zealand and How New Zealand Shaped Religion*. Palmerston North: Massey University Press.

Lonergan G, Lewis H, Tomalin E, and Waite L (forthcoming) Professionalisation or Secularisation? Understanding Postsecular Faith-based Anti-trafficking Responses. *Current Sociology*.

Lordon F (2014) *Willing Slaves of Capital: Spinoza and Marx on Desire*, trans. Ash G. London: Verso ebook version.

Luckmann T (1967) *The Invisible Religion: The Problem of Religion in Modern Society*. New York: The Macmillan Company.

Luz N (2013) Metaphors to Live by: Identity Formation and Resistance Among Minority Muslims in Israel. In Hopkins P, Kong L, and Olson E (eds) *Religion and Place: Landscape, Politics, and Piety*. New York, NY: Springer, pp. 57–74.

Lyons S (2014) The Disenchantment/Re-enchantment of the World. *Modern Language Review*, 109, 873–885.

MacDonald F (2002) Towards a Spatial Theory of Worship: Some Observations from Presbyterian Scotland. *Social & Cultural Geography*, 3(1), 61–80.

Madan TN (1997) *Modern Myths, Locked Minds: Secularism and Fundamentalism in India*. New Delhi: Oxford University Press.

Madan TN (2011) *Sociological Traditions: Methods and Perspectives in the Sociology of India*. New Delhi: Sage.

Madsen R (1998) *China's Catholics: Tragedy and Hope in an Emerging Civil Society*. California: Berkeley University of California.

Mahmood S (2005) *The Politics of Piety: The Islamic Revival and the Feminist Subject*. Princeton, NJ: Princeton University Press.

Mahmood S (2009) Religious Reason and Secular Affect: An Incommensurable Divide? *Critical Inquiry*, 35(4), 836–862.

Malik AR (2007) Take Me to Your Leader: Post-secular Society and the Islam Industry. *Eurozine* 23 April.

Malpass A, Cloke P, Barnett C, and Clarke N (2007) Fairtrade Urbanism? The Politics of Place Beyond Place in the Bristol Fairtrade City Campaign. *International Journal of Urban and Regional Research*, 31, 653–645.

Maoz I (2004) Peace Building in Violent Conflict: Israeli–Palestinian Post-Oslo People-to-People Activities. *International Journal of Politics, Culture, and Society*, 17(3), 563–574.

Marsh C (2003) *The Beloved Community: How Faith Shapes Social Justice, From the Civil Rights Movement to Today*. New York: Basic Books.

Martin D (1969) Notes for a General Theory of Secularisation. *European Journal of Sociology*, 10(2), 192–201.

Martin D (2011) *The Future of Christianity*. London: Routledge.

Martinson M (2013) Cultural Materiality and Spiritual Alienation. *Political Theology*, 14(2), 219–234.

Massey D (2004) Geographies of Responsibility. *Geografiska Annaler B*, 86, 5–18.

Massey D (2008) Geographies of Solidarities. In Clark N, Massey D, and Sarre P (eds) *Material Geographies: A World in the Making*. London: SAGE/Open University Press, pp. 311–362.

Massumi B (1997) The Political Economy of Belonging and the Logic of Relation. In Davidson C (ed.) *Anybody*. Cambridge, MA: MIT Press, pp. 224–238.

Mavelli L (2012) Postsecular Resistance, the Body, and the 2011 Egyptian Revolution. *Review of International Studies*, 38(5), 1057–1078.

Mavelli L and Petito F (2012) The Postsecular in International Relations: An Overview. *Review of International Studies*, 38(5), 931–942.

Mavelli L and Wilson E (2016) Postsecularism and International Relations. In Haynes J (ed.) *Routledge Handbook of Religion and Politics*. New York and London: Routledge, pp. 251–269.

May G (2018) Can Radical Hospitality be too Radical? Available from https://radio public.com/nomad-podcast-8Q20Pr/ep/s1!ae8. (last accessed 07/06/18).

May J and Cloke P (2014) Modes of Attentiveness: Reading for Difference in Geographies of Homelessness. *Antipode*, 46, 894–920.

May J, Cloke P, and Johnsen S (2006) Shelter at the Margins: New Labour and the Changing State of Emergency Accommodation for Single Homeless People in Britain. *Policy and Politics*, 34, 711–730.

McBroome K (2013) Revolutionaries: The Women of the Egyptian Uprising. Pace University (unpublished thesis). Available from www.pace.edu/sites/default/files/files/thesis-kerry-mcbroome.pdf (last accessed 25/05/18).

McCarraher E (2005) The Enchantments of Mammon: Notes Towards a Theological History of Capitalism. *Modern Theology*, 21(3), 429–461.

McClish C (2009) Activism Based in Embarrassment: The Anti-consumption Spirituality of the Reverend Billy. *Liminalities*, 5(2), 1–20.

McConnell M (2007) Secular Reason and the Misguided Attempt to Exclude Religious Argument from Democratic Deliberation. *Journal of Law, Philosophy and Culture*, 1, 159–174.

McCormack DP (2014) *Refrains for Moving Bodies: Experience and Experiment in Affective Spaces*. Durham, NC: Duke University Press.

McCowan T (2017) Building Bridges Rather than Walls: Research into an Experiential Model of Interfaith Education in Secondary Schools. *British Journal of Religious Education*, 39(3), 269–278.

McDonald C and Marston G (2005) Workfare as Welfare: Governing Unemployment in the Advanced Liberal State. *Critical Social Policy*, 25, 374–400.

McDonald K (2002) From Solidarity to Fluidarity: Social Movements Beyond 'Collective Identity' – The Case of Globalization Conflicts. Social Movement Studies. *Journal of Social, Cultural and Political Protest*, 1(2), 109–128.

McFarlane C (2011) *Learning the City: Knowledge and Translocal Assemblage*. Oxford: Wiley-Blackwell.

McGlynn C, Niens U, Cairns E, and Hewstone M (2004) Moving out of Conflict: The Contribution of Integrated Schools in Northern Ireland to Identity, Attitudes, Forgiveness and Reconciliation. *Journal of Peace Education*, 1(2), 147–163.

McIntosh A and Carmichael M (2016) *Spiritual Activism: Leadership as Service*. Cambridge: Green Books.

McKanan D (2011) *Prophetic Encounters: Religion and the American Radical Tradition*. Boston, MA: Beacon.

McLennan G (2007) Towards Postsecular Sociology? *Sociology*, 41, 857–870.

McLennan G (2010) The Postsecular Turn. *Theory, Culture and Society*, 27, 3–20.

McLennan G (2011) Postsecular Cities and Radical Critique: A Philosophical Seachange? In Beaumont J and Baker C (eds) *Postsecular Cities: space, theory and practice*. London: Continuum, pp. 15–30.

Megoran N (2010) Towards a Geography of Peace: Pacific Geopolitics and Evangelical Christian Crusade Apologies. *Transactions of the Institute of British Geographers*, 35(3), 382–398.

Megoran N (2013) Radical Politics and the Apocalypse: Activist Readings of Revelation. *Area*, 45(2), 141–147.

Melucci A (1996) *Challenging Codes: Collective Action in the Information Age*. Cambridge: Cambridge University Press.

Mendieta E (2010) A Postsecular World Society? On the Philosophical Significance of Postsecular Consciousness and the Multicultural World Society. Interview with Jurgen Habermas. *Monthly Review*, 21 March. Available from http://mrzine.monthlyreview.org/21010/habermas210310.html (last accessed 06/06/17).

Menon N (2007) Living with Secularism. In Needham A and Rajan R (eds) *The Crisis of Secularism in India*. Durham, NC: Duke University Press.

Mentinis M (2014) Towards a Revolutionary Psychology: On the Vicennial of the Zapatista Insurrection. *The Occupied Times*. Available from http://theoccupiedtimes.org/?p=12837 (last accessed 12/04/18).

Meyer B (2013) Lessons from 'Global Prayers': How Religion Takes Places in the City. In Becker J, Klingan K, Lanz S, and Wildner K (eds) *Global Prayers: Contemporary Manifestations of the Religious in the City*. Baden: Lars Muller Publishers, pp. 590–601.

Meyer F and Miggelbrink J (2017) Post-secular Rapprochement in Peripheralized Regions – Politics of Withdrawal and Parish Community Responses. *Geogr. Helv*, 72, 361–370.

Milbank J (1990) *Theology and Social Theory: Beyond Secular Reason*. Malden, MA: Blackwell.

Milbank J (1995) Only Theology Overcomes Metaphysics. *New Blackfriars*, 76(895), 325–343.

Milbank J (2005) Materialism and Transcendence. In Davis C, Milbank J, and Zizek S (eds) *Theology and the Political: New Debates*. Durham, NC: Duke University Press, pp. 393–426.

Milbank J (2006) *Theology and Social Theory: Beyond Secular Reason*. Oxford: Wiley-Blackwell, 2nd Edn.

Milbank J (N.D.) What is Radical Orthodoxy? Available from www.unifr.ch/theo/assets/files/SA2015/Theses_EN.pdf (last accessed 20/09/17).

Milbank J, Pickstock C, and Ward G (eds) (1999) *Radical Orthodoxy*. London: Routledge.

Mindock C (2017) Trump's Travel Ban on Six Muslim-majority Countries to be Fully Enacted after Supreme Court Ruling. Available from www.independent.co.uk/news/world/america/us-politics/trump-travel-ban-muslim-countries-supreme-court-ruling-allowed-go-ahead-latest-a8092086.html. (last accessed 15/05/2018).

Mitchell B (2017) *Faith Based Development*. Maryknoll, NY: Orbis.

Mitchell D (1997) The Annihilation of Space by Law: The Roots and Implications of Anti-homeless Laws in the United States. *Antipode*, 29(3), 303–335.

Modood T (2005) *Multicultural Politics: Racism, Ethnicity and Muslims in Britain.* Edinburgh: Edinburgh University Press.

Moghadam V (1995) Gender and Revolutionary Transformation: Iran 1979 and East Central Europe 1989. *Gender and Society*, 9, 328–358.

Molendijk A, Beaumont J, and Jedan C (eds) (2010) *Exploring the Postsecular: The Religious, the Political and the Urban.* Leiden: Brill.

Monaghan M (2012) The Recent Evolution of UK Drug Strategies: From Maintenance to Behaviour Change? *People, Place & Policy Online*, 6, 29–40.

Monbiot G (2017) *Out of the Wreckage: A New Politics for an Age of Crisis.* New York: Verso Books.

Moody KS (2010) 'I Hate Your Church; What I Want is My Kingdom': Emerging Spiritualities in the UK Emerging Church Milieu. *The Expository Times*, 121(10), 405–503.

Moody KS (2012) Retrospective Speculative Philosophy: Looking for Traces of Žižek's Communist Collective in Emerging Christian Praxis. *Political Theology*, 13(2), 183–199.

Muehlebach A (2012) *The Moral Neoliberal: Welfare and Citizenship in Italy.* Chicago: University of Chicago Press.

Mufti A (2013) Part 1: Why I Am Not a Postsecularist. *Boundary 2: An International Journal of Literature and Culture*, 40(1), 7–19.

Murphy S (2009) 'Compassionate' Strategies of Managing Homelessness: Post-revanchist Geographies in San Francisco. *Antipode*, 41(2), 305–325.

Murray S (2011) *Post-Christendom: Church and Mission in a Strange New World.* Milton Keynes: Paternoster Press.

Nancy JL (1994) Cut Throat Sun. In Arteaga A (ed.) *An Other Tongue: Nation and Ethnicity in the Linguistic Borderlands.* Durham, NC: Duke University Press, p. 113.

Nandy A (1997) Twilight of Certitudes: Secularism, Hindu Nationalism and Other Masks of Deculturation. *Alternatives*, 22(2), 157–176.

Narayanan Y (ed.) *Religion and Urbanism: Reconceptualising Sustainable Cities for South Asia.* London and New York: Routledge.

Ng and Fulda (2017) Religious Dimensions of Hong Kong's Umbrella Movement. *Journal of Church and State*, 1–21. https://doi.org/10.1093/jcs/csx053.

Nietzsche F (1974) *The Gay Science*, trans. Kauffmann W. New York, NY: Vintage.

Nita M (2014) Christian and Muslim Climate Activists Fasting and Praying for the Planet: Emotional Translation of 'Dark Green' Activism and Green-Faith Identities. In Globus-Veldman R, Szasz A, and Haluza-DeLay R (eds) *How the World's Religions are Responding to Climate Change Social Scientific Investigation.* New York: Routledge, pp. 229–243.

Norris P and Inglehart R (2004) *Sacred and Secular: Religion and Politics Worldwide.* Cambridge: Cambridge University Press.

Noxolo P, Raghuram P, and Madge C (2012) Unsettling Responsibility: Postcolonial Interventions. *Transactions of the Institute of British Geographers*, 37, 418–429.

Nusseibeh L (2011) Women and Power in the Israeli–Palestinian Conflict. *Palestine–Israel: A Journal of Politics, Economics and Culture*, 17(3–4). Available from www.pij.org/details.php?id=1371 (last accessed 27/07/18).

Nynas P, Lassander M, and Utriainen T (eds) (2015) *Post-secular Society.* New Brunswick, NJ: Transaction.

O'Neill K (2014) On Liberation: Crack, Christianity, and Captivity in Postwar Guatemala City. *Social Text*, 32, 11–28.

O'Neill S (2000) The Politics of Inclusive Agreements: Towards a Critical Discourse Theory of Democracy. *Political Studies*, 48, 503–521.

Olson E (2016) Geography and Ethics II: Emotions and Morality. *Progress in Human Geography*, 40(6), 830–838.

Olson E, Hopkins P, Pain R, and Vincent G (2013) Retheorizing the Postsecular Present: Embodiment, Spatial Transcendence, and Challenges to Authenticity Among Young Christians in Glasgow, Scotland. *Annals of the Association of American Geographers*, 103, 1421–1436.

Olson K (2005) Music for Community Education and Emancipatory Learning. *New Directions for Adult and Continuing Education*, 107(Autumn), 55–64.

Ong A (2007) Neoliberalism as a Mobile Technology. *Transaction of the Institute of British Geographers*, 32(1), 3–8.

Osuri G (2012) (Post) Secular Discomforts: Religio-Secular Disclosures in the Indian Context. *Cultural Studies Review*, 18, 32–51.

Özyürek E (2006) *The Nostalgia for the Modern: State Secularism and Everyday Politics in Turkey*. Raleigh: Duke University Press.

Pacione M (1999) The Relevance of Religion for a Relevant Human Geography. *Scottish Geographical Journal*, 115, 117–131.

Parish S (1997) Goddesses Dancing in the City: Hinduism in an Urban Incarnation – A Review Article. *International Journal of Hindu Studies*, 1, 441–484.

Parsons M (2014) *Rubble to Resurrection: Churches Respond in the Canterbury Quakes*. Auckland: Daystar.

Phelps H (2013) Resonating Moral Monday. *Political Theology Network* 11 July 2013. Available from www.politicaltheology.com/blog/resonating-moral-monday/ (last accessed 22/07/18).

Pickerill J and Chatterton P (2006) Notes Towards Autonomous Geographies: Creation, Resistance and Self-management as Survival tactics. *Progress in Human Geography*, 30(6), 730–746.

Pickles K (2016) *Christchurch Ruptures*. Auckland: Bridget Williams Books.

Pierson M (2012) *The Art of Curating Worship*. London: Canterbury Press Norwich.

Plender A (2018) After Grenfell: The Faith Groups' Response. Theos Report: London. Available from www.theosthinktank.co.uk/research/2018/06/01/after-grenfell-the-faith-groups-response (last accessed 20/07/18).

Pollard J and Samers M (2007) Islamic Banking and Finance: Postcolonial Political Economy and the Decentring of Economic Geography. *Transactions of the Institute of British Geographers*, 32, 313–330.

Polletta F and Jasper JM (2001) Collective Identity and Social Movements. *Annual Review of Sociology*, 27, 283–305.

Popke J (2007) Geography and Ethics: Spaces of Cosmopolitan Responsibility. *Progress in Human Geography*, 31, 509–518.

Popke J (2009) Ethical Spaces of Being In-common. In Smith S, Pain R, Marston S, and Jones III J-P (eds) *Handbook of Social Geography*. London: Sage, pp. 435–454.

Possamai A (2017) *The i-zation of Society, Religion, and Neoliberal Post-Secularisation*. Basingstoke: Palgrave Macmillan.

Potter P (2003) Belief in Control: Regulation of Religion in China. *The China Quarterly*, 174, 338–358.

Powell H (2004) A Dream Wedding: From Community Music to Music Therapy with a Community. In Pavlicevic M and Ansdell G (eds) *Community Music Therapy*. London: Jessica Kingsley Publishers, pp. 167–185.

Prior J and Crofts P (2015) Shooting Up Illicit Drugs with God and the State: The Legal-spatial Constitution of Sydney's Medically Supervised Injecting Centre as a Sanctuary. *Geographical Research*, 54(3), 313–323.

Prochaska F (2006) *Christianity and Social Service in Modern Britain*. Oxford: Oxford University Press.

Qian J and Kong L (2018) Buddhism Co. Ltd? Epistemology of Religiosity, and the Reinvention of a Buddhist Monastery in Hong Kong. *Environment and Planning D Society and Space*, 36, 159–177.

Radden J (ed.) (2000) *The Nature of Melancholy: From Aristotle to Kristeva*. Oxford: Oxford University Press.

Rashbrooke M (ed.) (2013) *Inequality: A NZ Crisis*. Auckland: Bridget Williams Books.

Rashkover R (2017) Covenantal Ethics and the 2016 Election. *Political Theology*, 18(3), 201–205.

Ratti M (2013) *The Postsecular Imagination: Postcolonialism, Religion and Literature*. London and New York: Routledge.

Rawls J (1995) Political Liberalism: A Reply to Habermas. *Journal of Philosophy*, 92, 132–180.

Rawls J (1997) The Idea of Public Reason Revisited. *The University of Chicago Law Review*, 64, 765–807.

Rawls J (2005) *Political Liberalism: Expanded Edition*. New York: Columbia University Press.

Reder M and Schmidt J (2010) Habermas and Religion. In Habermas J, Brieskorn N, Reder M, Ricken F, and Schmidt J (eds) *An Awareness of What is Missing: Faith and Reason in a Post-Secular Age*. Cambridge: Polity Press, pp. 1–14.

Reinhard K (2005) Toward a Political Theory of the Neighbour. In Zizek S, Santner E, and Reinhard K (eds) *The Neighbor: Three Inquiries on Political Theology*. Chicago, IL: University of Chicago Press, pp. 11–75.

Reuver M (1988) *Christians as Peacemakers: Peace Movements in Europe and the USA*. Geneva: WCC Publications.

Reynolds R (2014) Desire for the Gap. In Bennett B, Dann J, Johnson E, and Reynolds R (eds) *Once In A Lifetime: City-building After Disaster in Christchurch*. Christchurch: Freerange Press, pp. 167–176.

Richardson E (2013) Using Performance in Human Geography: Conditions and Possibilities. *Kaleidoscope: The Interdisciplinary Postgraduate Journal of Durham University's Institute of Advanced Study*, 5(1), 124–133.

Rieger J (2001) *God and the Excluded: Visions and Blindspots in Contemporary Theology*. Minneapolis: Fortress.

Rieger J (2012) Power or Glory? Available from http://aeon.co/magazine/society/joerg-rieger-occupy-religion/ (last accessed 28/08/15).

Rieger J and Pui-lan K (2012) *Occupy Religion: Theology of the Multitude*. Plymouth: Rowman & Littlefield Publishers, Inc.

Riesebrodt M (2014) *Religion in the Modern World: Between Secularization and Resurgence*. European University Institute Max Weber Programme. Available from http://cadmus.eui.eu/bitstream/handle/1814/29698/MWP_LS_2014_01_Riesebrodt.pdf (last accessed 26/05/18).

Roberts JD (2005) *Bonhoeffer & King: Speaking Truth to Power*. Louisville, KY: Westminster John Knox Press.

Rodriguez T (2010) Bio-pistis: Conversion of Heroin Addicts in Prisons, On Medicine, and with God. In Adkins J, Occhipinti L, and Hefferan T (eds) *Not by Faith Alone: Social Services, Social Justice, and Faith-based Organizations in the United States*. Lanham, MD: Lexington Books, Rowman & Littlefield, pp. 207–230.

Roe G (2005) Harm Reduction as Paradigm: Is Better Than Bad Good Enough? The Origins of Harm Reduction. *Critical Public Health*, 15(3), 243–250.

Rohr R (2003) *Everything Belongs: The Gift of Contemplative Prayer*. New York, NY: The Crossroad Publishing Company.

Rollins P (2006) *How (Not) to Speak of God*. London: SPCK.

Romanillos J, Beaumont J, and Sen M (2012) State-religion Relations and Welfare Regimes in Europe. In Beaumont J and Cloke P (eds) *Faith-based Organizations and Exclusion in European Cities*. Bristol: Policy Press, pp. 37–58.

Roome D (2012) *Christchurch Earthquake Images*. Wellington: Awa Press.

Rose G (1997) Situating Knowledges: Positionality, Reflexivities and Other Tactics. *Progress in Human Geography*, 21(3), 305–320.

Rose M (2017) Machines of Loving Grace: Angels, Cyborgs, and Postsecular Labour. *Journal for Cultural and Religious Theory*, 16(2), 240–259.

Routledge P (2003) Convergence Space: Process Geographies of Grassroots Globalization Networks. *Transactions of the Institute of British Geographers*, 28(3), 333–349.

Ruud E (2004) Foreword: Reclaiming Music. In Pavlicevic M and Ansdell G (eds) *Community Music Therapy*. London: Jessica Kingsley Publishers, pp. 11–14.

Rygiel K (2011) Bordering Solidarities: Migrant Activism and the Politics of Movement and Camps at Calais. *Citizenship Studies*, 15(1), 1–19.

Saler A (2011) *As If: Modern Enchantment and the Literary Prehistory of Virtual History*. Oxford: Oxford University Press.

Salter M (2013) Carl Schmitt on the Secularisation of Religious Texts as a Resacralisation of Jurisprudence? *International Journal for the Semiotics of Law*, 26, 113–147.

Sandel M (2012) *What Money Can't Buy: The Moral Limits of Markets*. London: Allen Lane.

Sandri E (2018) 'Volunteer Humanitarianism': Volunteers and Humanitarian Aid in the Jungle Refugee Camp of Calais. *Journal of Ethnic and Migration Studies*, 44(1), 65–80.

Sartre JP (2003) *Being and Nothingness: An Essay on Phenomenological Ontology*. London: Routledge.

Sayer A (2015) *Why We Can't Afford the Rich*. Bristol: Policy Press.

Schaper D (2016) Sanctuary Movement Sees Post-election Resurgence. Here's How to Get Involved. Available from https://sojo.net/articles/sanctuary-movement-sees-post-election-resurgence-heres-how-get-involved (last accessed 15/05/18).

Schmitt C (2005) [1922] *Schmitt, Political Theology, Four Chapters on the Concept of Sovereignty*, trans. Schwab G. Chicago: University of Chicago Press.

Schumaker J (2016) The Demoralized Mind. *The New Internationalist*. April 2016. Available from https://newint.org/columns/essays/2016/04/01/psycho-spiritual-crisis (last accessed 10/11/17).

Sen A (2005) Secularism and Its Discontents. *The Argumentative Indian: Writings on Indian History, Culture and Identity*. London: Penguin Books, pp. 294–316.

Sherwood H (2016) Hundreds of Churches Offer Sanctuary to Undocumented Migrants after Election. *Guardian*. 27 November 2016. Available from www.theguardian.com/us-news/2016/nov/27/undocumented-immigrations-us-churches-sanctuary-trump?CMP=share_btn_link (last accessed 15/05/18).

Sherwood H (2017) Grenfell: Faith Groups Step in to Mediate Between Officials and

Community. *Guardian.* 19 July 2017. Available from www.theguardian.com/uk-news/2017/jul/19/grenfell-faith-groups-step-in-to-mediate-between-officials-and-community (last accessed 21/07/18).

Shildrick T (2018) Lessons from Grenfell: Poverty Propaganda, Stigma and Class Power. *The Sociological Review Monographs,* 66(4), 783–798.

Shortall H (2015) Nine Things I Learned Volunteering in a Refugee Camp. 13 October 2015. Available from https://hollyshortall.wordpress.com/2015/10/13/nine-things-i-learned-volunteering-in-a-refugee-camp/ (last accessed 21/07/18).

Sibley C and Bulbulia J (2012) Faith after an Earthquake. A Longitudinal Study of Religion and Perceived Health Before and After the 2011 Christchurch New Zealand Earthquake. *PLoS ONE,* 7(12): e49648. Available from https://doi.org/10.1371/journal.pone.0049648 (last accessed 29/03/18).

Simmel G (2002) [1903] The Metropolis and Mental Life. In Bridge G and Watson S (eds) *The Blackwell City Reader.* Oxford and Malden, MA: Wiley-Blackwell, pp. 11–19.

Simpson P (2017) Spacing the Subject: Thinking Subjectivity after Non-representational Theory. *Geography Compass,* 10.1111/gec3.12347.

Sitrin M (2006) *Horizontalism: Voices of Popular Power in Argentina.* Oakland: AK Press.

Sitrin M (2012) *Everyday Revolutions: Horizontalism and Autonomy in Argentina.* New York and London: Zed.

Slater T (2014) The Myth of 'Broken Britain': Welfare Reform and the Production of Ignorance. *Antipode,* 46(4), 948–969.

Slessarev-Jamir H (2011) *Prophetic Activism: Progressive Religious Justice Movements in Contemporary America.* New York and London: New York University Press.

Smith C (1996) *Disruptive Religion: The Force of Faith in Social Movement Activism.* London: Routledge.

Smith C (2016) *Addiction, Modernity, and the City: A Users' Guide to Urban Space.* London: Routledge.

Smith D (2000) *Moral Geographies: Ethics in a World of Difference.* Edinburgh: Edinburgh University Press.

Smith N (1992) Contours of a Spatialized Politics: Homeless Vehicles and the Production of Geographical Scale. *Social Text,* 33, 54–81.

Smith TA, Murrey A, and Leck H (2017) 'What Kind of Witchcraft is This?' Development, Magic and Spiritual Ontologies. *Third World Thematics,* 2(2–3), 141–156.

Snow DA and Anderson L (1993) *Down on Their Luck: A Study of Homeless Street People.* Berkeley: University of California Press.

Sojourners Website. We the People Cannot Be Silent. Available from https://sojo.net/media/we-people-cannot-be-silent (last accessed 15/05/18).

Solnit R (2009) *A Paradise Built in Hell: The Extraordinary Communities that Arise in Disaster.* New York: Viking Penguin.

Southern N (2011) Strong Religion and Political Viewpoints in a Deeply Divided Society: An Examination of the Gospel Hall Tradition in Northern Ireland. *Journal of Contemporary Religion,* 26(3), 433–449.

Sparke M (2013) From Global Dispossession to Local Repossession: Towards a Worldly Cultural Geography of Occupy Activism. In Johnson N, Schein R, and Winders J (eds) *Companion to Cultural Geography.* Oxford: Wiley-Blackwell, pp. 167–185.

Spivak G (2004) Terror: A Speech After 9–11. *Boundary,* 31, 81–111.

Squire V (2011) From Community Cohesion to Mobile Solidarities: The City of Sanctuary Network and the Strangers into Citizens Campaign. *Political Studies,* 59, 290–307.

Stacey T (2017) Imagining Solidarity in the Twenty-first Century: Towards a Performative Postsecularism. *Religion, State and Society*, 45(2), 141–158.

Staeheli LA (2013) THE 2011 ANTIPODE AAG LECTURE – Whose Responsibility Is It? Obligation, Citizenship and Social Welfare. *Antipode*, 45(3), 521–540.

Stark R and Finke R (2000) *Acts of Faith: Explaining the Human Side of Religion.* Berkeley: University of California Press.

Stepan A (2011) The Multiple Secularisms of Modern Democratic and Non-Democratic Regimes. In Calhoun C, Juergensmeyer M, and VanAntwerpen J (eds) *Rethinking Secularism.* Oxford: Oxford University Press.

Storper M and Scott A (2016) Current Debates in Urban Theory: A Critical Assessment. *Urban Studies*, 53, 1114–1136.

Stout J (2004) *Democracy and Tradition.* Princeton, NJ: Princeton University Press.

Sullivan S (2005) An Other World is Possible? On Representation, Rationalism and Romanticism in Social Forum. *Ephemera*, 5(2), 370–392.

Suter K (2008) The Future of the Nation-state in an Era of Globalization. *Medicine, Conflict and Survival*, 24, 201–218.

Sutherland C (2014) Political Discourse and Praxis in the Glasgow Church. *Political Geography*, 38, 23–32.

Sutherland C (2016) Theography and Postsecular Politics in the Geographies of Postchristendom Communities (Doctoral Thesis). University of Exeter. Available from http://ore.exeter.ac.uk/repository/handle/10871/27782.

Sutherland C (2017) Theography: Subject, Theology, and Praxis in Geographies of Religion. *Progress in Human Geography*, 41(3), 321–337.

Sweeney J (2008) Revising Secularization Theory. In Ward G and Hoelzl M (eds) *The New Visibility of Religion.* London: Continuum, pp. 15–29.

Talvacchia KT, Pettinger MF, and Larrimore M (eds) (2014) *Queer Christianities: Lived Religion in Transgressive Forms.* New York, NY: New York University Press.

Tarlo E (2007) Hijab in London: Metamorphosis, Resonance and Effects. *Journal of Material Culture*, 12, 131–156.

Taubes J (2004) *The Political Theology of Saint Paul.* Stanford, CA: Stanford University Press.

Taylor B (2010) *Dark Green Religion Nature: Spirituality and the Planetary Future.* Berkeley and London: University of California Press.

Taylor C (2002) *Varieties of Religion Today.* Cambridge, MA: Harvard University Press.

Taylor C (2007) *A Secular Age.* Cambridge, MA: Harvard University Press.

Taylor C (2010) The Meaning of Secularism. *Hedgehog Review*, Fall, 23–34.

Taylor C (2011) Why We Need a Radical Redefinition of Secularism. In Butler J, Habermas J, Taylor C, West C, edited by Mendieta E and VanAntwerpen J, afterword by Calhoun C. *The Power of Religion in the Public Sphere.* New York: Columbia University Press, pp. 34–59.

Taylor MC (1987) *Erring: A Postmodern A/theology.* Chicago, IL: University of Chicago Press.

Tharoor I (2016) ISIS Calls for Holy War Find an Echo in Pro-Trump Movement. *Washington Post* 16 November 2016. Available from www.washingtonpost.com/news/worldviews/wp/2016/11/16/isis-wants-to-fight-a-holy-war-so-do-some-trump-supporters/?utm_term=.43b3b2b7eb3e (last accessed 20/07/18).

The Holy Bible: English Standard Version (2001) Wheaton, IL: Crossway.

Thomas S (2013) Re-engaging with the Margins: The Salvation Army 614 UK Network and Incarnational Praxis. In Cloke P, Beaumont J and Williams W (eds) *Working*

Faith: Faith-based Organisations and Urban Social Justice. Milton Keynes: Paternoster Press, pp. 66–84.

Thrift N (1996) *Spatial Formations.* London: Sage.

Tosi S and Vitale T (2009) Explaining How Political Culture Changes: Catholic Activism and the Secular Left in Italian Peace Movements. *Social Movement Studies: Journal of Social, Cultural and Political Protest,* 8(2), 131–147.

Tse JKH (2014) Grounded Theologies: 'Religion' and the 'Secular' in Human Geography. *Progress in Human Geography,* 38, 201–220.

Tse JKH (2015) Under the Umbrella: Grounded Christian Theologies and Democratic Working Alliances in Hong Kong. *Review of Religion and Chinese Society,* 2(1), 109–142.

Tse JKH and Tan J (ed.) (2016) *Theological Reflections on the Hong Kong Umbrella Movement.* New York: Palgrave.

Turner V (ed.) (1969) Liminality and Communitas. *The Ritual Process: Structure and Anti-Structure.* Chicago: Aldine Publishing Company, pp. 358–374.

Tyler I (2017) *Revolting Subjects: Social Abjection and Resistance in Neoliberal Britain.* London: Zed Books.

Valentine G (2008) Living with Difference: Reflections on Geographies of Encounter. *Progress in Human Geography,* 32(3), 323–337.

Valentine G and Sadgrove J (2012) Lived Difference: A Narrative Account of Spatiotemporal Processes of Social Differentiation. *Environment and Planning A,* 44, 2049–2063.

Valentine G and Waite L (2011) Negotiating Difference through Everyday Encounters: The Case of Sexual Orientation and Religion and Belief. *Antipode,* 44, 474–492.

Valverde M (1998) *Diseases of the Will: Alcohol and the Dilemma of Freedom.* Cambridge: Cambridge University Press.

Van Steenwyk M (2013) *The Unkingdom of God: Embracing the Subversive Power of Repentance.* Downers Grove, IL: InterVarsity Press.

Vasiliki R (2016) The Politics of Postsecular Feminism. *Theory, Culture and Society,* 33, 103–123.

Vattimo G (2002) *After Christianity.* New York: Columbia University Press.

Vattimo G (2003) After Onto-theology: Philosophy Between Science and Religion. In Wrathall M (ed.) *Religion After Metaphysics.* Cambridge: Cambridge University Press, pp. 29–36.

Vattimo G (2007) Towards a Nonreligious Christianity. In Caputo J and Vattimo G (eds) *After the Death of God.* New York: Columbia University Press, pp. 27–46.

Vattimo G (2009) *Not Being God: A Collaborative Autobiography.* New York: Columbia University Press.

Vincett G (2013) 'There's Just No Space for Me There': Christian Feminists in the UK and the Performance of Space and Religion. In Hopkins P, Kong L, and Olson E (eds) *Religion and Place: Landscape, Politics and Piety.* New York, NY: Springer, pp. 167–184.

Voinea A (2014) The Co-operatives Acting as Agents for Peace. *Cooperative News* 13 August 2014. Available from www.thenews.coop/88497/topic/development/the-co-operatives-acting-as-agents-for-peace/ (last accessed 27/07/18).

Volf M (1996) *Exclusion and Embrace: A Theological Exploration of Identity, Otherness and Reconciliation.* Nashville, TN: Abingdon Press.

Vrasti W (2009) How to use Affective Competencies in Late Capitalism. Paper presented in British International Studies Association Conference, University of Leicester, December 2009. Available from www.bisa.ac.uk/index.php?option=com_bisa&task=download_paper&no_html=1&passed_paper_id=29 (last accessed 20/07/18).

Walker P (2001) *Pulling the Devil's Kingdom Down: The Salvation Army in Victorian Britain.* London: University of California Press.

Walters W (2008) Acts of Demonstration: Mapping the Territory of (Non-)Citizenship. In Isin EF and Nielsen GM (eds) *Acts of Citizenship.* London: Zed Books, pp. 182–206.

Ward G (2000) *Cities of God.* London: Routledge.

Ward G (2005) *Cultural Transformation and Religious Practice.* Cambridge: Cambridge University Press.

Ward G (2009) *The Politics of Discipleship: Becoming Postmaterial Citizens.* Grand Rapids, MI: Baker Academic.

Ward G (2014) The Myth of Secularism. *Telos,* 167, 162–179.

Washington Post (2017) Deconstructing the symbols and slogans spotted in Charlottesville. *Washington Post* 18 August 2017. Available from www.washingtonpost.com/graphics/2017/local/charlottesville-videos/?utm_term=.c7ca4f3bf56b (last accessed 20/07/18).

Watson J (2013) Post-secular Schooling: Freedom Through Faith or Diversity in Community. *Cambridge Journal of Education,* 43(2), 147–162.

Weber M (1976) *The Protestant Ethic and the Spirit of Capitalism.* London: Allen and Unwin, 2nd Edn.

White M (2016) *The End of Protest: A New Playbook for Revolution.* Toronto: Penguin Random House.

White M (2018) Micah White's Thought Bubble: Junk Thought. Available from www.micahwhite.com/micah-whites-though-bubble/ (last accessed 16/05/18).

Whitehead A, Perry S, and Baker J (2018) Make America Christian Again: Christian Nationalism and Voting for Donald Trump in the 2016 Presidential Election. *Sociology of Religion,* 79(2), 147–171.

Wickström L and Illman R (2015) Environmentalism as a Trend in Post-Secular Society. In Nynäs P, Lassander M, and Utrianinen T (eds) *Post-Secular Society.* New Brunswick: Transaction Publishers, pp. 217–238.

Wilford J (2010) Sacred Archipelagos: Geographies of Secularization. *Progress in Human Geography,* 34, 328–348.

Williams A (2013) Practical Theology and Christian Responses to Drug Use. In Cloke P, Beaumont J, and Williams A (eds) *Working Faith: Faith-based Organizations and Urban Social Justice.* Milton Keynes: Paternoster Press, pp. 47–65.

Williams A (2015) Postsecular Geographies: Theo-ethics, Rapprochement and Neoliberal Governance in a Faith-based Drug Programme. *Transactions, Institute of British Geographers,* 40, 192–208.

Williams A (2017) Residential Ethnography, Mixed Loyalties, and Religious Power: Ethical Dilemmas in Faith-based Addiction Treatment. *Social & Cultural Geography,* 18(7), 1016–1038.

Williams A, Cloke P, May J, and Goodwin M (2016) Contested Space: The Contradictory Political Dynamics of Food Banking in the UK. *Environment and Planning A,* 48(11), 2291–2316.

Williams A, Cloke P, and Thomas S (2012) Co-constituting Neoliberalism: Faith-based Organisations, Co-option, and Resistance in the UK. *Environment and Planning A,* 44(6), 1479–1501.

Williams A, May J, Cloke P, and Cherry L (forthcoming) *Feeding Austerity? Ethical Ambiguity and Political Possibilities in UK Foodbanks.* Oxford: Wiley-Blackwell.

Williams DS (2013) *Sisters in the Wilderness: The Challenge of Womanist God-Talk.* New York, NY: Orbis Books.

Williams P (2015) *Everyday Peace? Politics, Citizenship and Muslim Lives in India.* Oxford: Wiley-Blackwell.

Willis AC (2017) Notes Toward a Dissident Theo-Politics. *Political Theology*, 18(4), 290–308.

Wilson E (2014) Theorizing Religion as Politics in Postsecular International Relations. *Politics, Religion & Ideology*, 15(3), 347–365.

Wilson HF (2014) Multicultural Learning: Parent Encounters with Difference in a Birmingham Primary School. *Transactions of the Institute of British Geographers*, 39(1), 102–114.

Wilson J and Swyngedouw E (2014) Seeds of Dystopia: Post-politics and the Return of the Political. In Wilson J and Swyngedouw E (eds) *The Post-Political and its Discontents.* Edinburgh: Edinburgh University Press, pp. 1–24.

Wilton R and DeVerteuil G (2006) Spaces of Sobriety/Sites of Power: Examining Social Model Alcohol Recovery Programs as Therapeutic Landscapes. *Social Science and Medicine*, 63, 649–661.

Winter E (2017) An Activist Religiosity? Exploring Christian Support for the Occupy Movement. *Journal of Contemporary Religion*, 32(1), 51–66.

Wolterstorff N (2007) The Paradoxical Role of Coercion in the Theory of Political Liberalism. *Journal of Law, Philosophy and Culture*, 1, 101–125.

Woodhead L (2012) Introduction. In Woodhead L and Catto R (eds) *Religion and Change in Modern Britain.* London: Routledge, pp. 1–33.

Woodward K (2011) Affective life. In Del Casino Jr VJ, Thomas ME, and Cloke P (eds) *A Companion to Social Geography.* Oxford: Wiley-Blackwell, pp. 326–345.

Yates GJ and Silverman MJ (2016) Needs of Children Experiencing Homelessness who are Living in Shelters: A Qualitative Investigation of Perceptions of Care Workers to Inform Music Therapy Clinical Practice. *Voices: A World Forum for Music Therapy*, 16(3), 77–101.

Yeoh B (1996) *Contesting Space in Colonial Singapore: Power Relations and the Urban Built Environment.* Singapore: Singapore University Press.

Yoder J (1994) *The Politics of Jesus.* Grand Rapids: Eerdmans.

Yorgason E and della Dora V (2009) Geography, Religion, and Emerging Paradigms: Problematizing the Dialogue. *Social & Cultural Geography*, 10(6), 629–637.

Young I (2004) Responsibility and Global Labor Justice. *Journal of Political Philosophy*, 12, 365–388.

Yukich G (2013) *One Family Under God: Immigration Politics and Progressive Religion in America.* Oxford University Press: New York.

Zhe J (2011) Buddhism in the Reform Era: A Secularized Revival? In Chau A (ed.) *Religion in Contemporary China: Revitalization and Innovation.* London and New York: Routledge.

Zimmerman Y (2011) Christianity and Human Trafficking. *Religion Compass*, 5(10), 567–578.

Žižek S (2000) *The Fragile Absolute: Or Why is the Christian Legacy Worth Fighting For?* London: Verso.

Žižek S (2001) *On Belief.* London: Routledge.

Žižek S (2004) *Plea for Ethical Violence.* Talk given to The European Graduate School. Zürich: EGS.

Žižek S (2011) *Living in the End Times.* London: Verso.

Žižek S (2017) The Courage of Hopelessness: Chronicles of a Year of Acting Dangerously. London: Allen Lane.

Index

238   *Index*